Artificial intelligence applied to solve social problems in the era of moderny and postmodernity

Alberto Ochoa Zezzatti

Nemesio Castillo Viveros

Centro de Investigaciones Sociales de la Universidad Autónoma de Ciudad Juárez

Printed in USA

Index

Aislamiento Social de familiares en primer grado sin descendencia directa

"Tú si me entiendes, tú si me comprendes, tú deberías de ser mi mamá"

Maricruz a su Tía Eduwiges en la telenovela "Quinceañera"

Alberto Ochoa-Zezzatti[1], Nemesio Castillo[1], Emmanuel García[1] & Ainhoa Kauranderaín[2]

Juarez City University[1], Basque Country University[2]

El propósito de la presente investigación es comprender desde una perspectiva Multivariable la situación existente en cuanto al aislamiento que sufre una minoría que aumenta en número desde finales del siglo pasado, como lo son las personas que deciden no tener un compañero sentimental, determinar las causas que originan este comportamiento social y las recomendaciones en cuanto a una política pública que debería de existir para poder ayudar a este grupo que para el 2035 representará el 28.7% de la población total en Ciudad Juárez.

Existen diversos factores que pueden incidir en cuanto a que una persona decida no casarse y permanecer soltera aún a una edad avanzada, estos factores incluyen desde la propia decisión hasta fracasos sentimentales de todo tipo y viudez.

Primeramente, en la sección 1 de nuestra investigación, explicamos el concepto de Aislamiento Social en especial el enfocado a esta minoría y los efectos a largo plazo en Ciudad Juárez. En la sección 2 se explica un estudio realizado con respecto a la percepción social que se tiene acerca de los problemas ocasionados por el abandono social esto visto desde la perspectiva de un grupo de control. En la Sección 3, describimos un análisis situacional, relacionado con una muestra con el apoyo de una Organización No Gubernamental. Finalmente en la Sección 4, realizamos un análisis Multivariable de la información obtenida utilizando Social Data Mining y detallamos nuestras consideraciones finales acerca de esta situación social que afecta las esperanzas, sueños e ilusiones de familiares en primer grado que apoyaron durante toda su vida a su familia, y que al final de la misma quedan en el abandono por una u otra razón por parte de sus propios familiares.

1.- Descripción de la Problemática.

Ciudad Juárez, la Ciudad más poblada en la frontera con Estados Unidos el estado más grande de la Federación Mexicana y ubicado en el Norte del país, es una Sociedad Multicultural al igual que las otras entidades federativas cercanas a la fronteras, desde mediados de los años 80's, oleadas temporales provenientes desde Zacatecas, Durango, Coahuila & Veracruz dan como resultado un mosaico diverso que debe de convivir no siempre en las mejores condiciones para desarrollarse en el entorno familiar, un aspecto importante a destacar es que el abandono social es alto en esta sociedad, pero afecta en una proporción más alta a personas adultas mayores de 55 años, esta investigación pretende explicar las razones que justificarían el cúmulo de factores que determinan la vulnerabilidad social de este grupo.

El aislamiento social, está determinado por un abandono por parte de sus familias relacionado con muchas actividades sociales, en el caso del grupo conformado por personas mayores de 55 años y que no tienen una pareja en cohabitación ni descendientes directos que puedan cuidar de ellos, el grupo es completamente heterogéneo desde personas que han decidido vivir esta situación como un estilo de vida hasta quienes cohabitan con alguno de sus padres o sus hermanos casados y sus familias.

Para poder determinar en una forma más objetiva, esta comparativa, decidimos utilizar Social Data Mining, en especial un análisis de Sociolingüística para poder evaluar los comentarios vertidos en una serie de entrevistas colectivas realizadas en Ciudad Juárez durante el primer semestre de 2012, y por otra parte una investigación de indicadores socioeconómicos para este grupo vulnerable, en la mayor parte de estos indicadores se describen de una manera informal, los pormenores del abandono social realizado por sus propias familias e incluso muchas veces se puede tener acceso a datos del tiempo que han permanecido solos, su ubicación geoespacial, situación socioeconómica y categoría social.

2.- Metodología utilizada

El nombre de Data Mining, esta relacionado con las similitudes entre buscar por información valiosa en grandes bases de datos – por ejemplo: encontrar información de las tendencias del comportamiento social en grandes cantidades de Gigabytes almacenados – y el minado de una montaña para encontrar una veta de metales valiosos. Minería de Datos automatiza el proceso de encontrar información predecible en grandes bases de datos (Ver Figura 1). Preguntas que tradicionalmente requieren un análisis manual intensivo ahora pueden ser directamente y fácilmente respondidas desde la información [1].

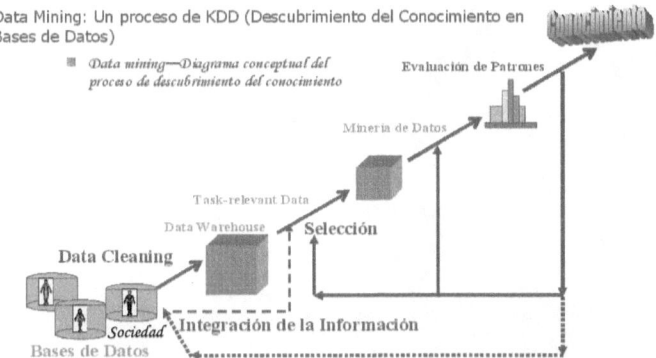

Figura 1. Proceso de Minería de Datos. La sociedad de la información se ubica dentro de *Bases de Datos*, las cuáles son limpiadas y almacenadas en *un Data Ware House*, entonces son minadas mediante un ciclo hacia atrás dentro de los procesos de selección y evaluación de patrones respectivamente.

El primer paso para desarrollar una adecuada proyección de predicción numérica basada en Minería da Datos, fue determinar la proporcionalidad que existe entre los diferentes grupos por categoría para Ciudad Juárez y poder determinar la cantidad de individuos en soltería dentro del rango de edad de 55 años y más, analizar mediante una proyección numérica el crecimiento que tendrá este grupo social, como lo demuestra la pirámide poblacional correspondiente a Ciudad Juárez, ver figura 2.

Uno de los aspectos más importantes que se pueden observar en Chihuahua es la diversidad en cuanto a patrones culturales establecidos por cada categoría, una perspectiva de ver la cotidianeidad desde un constructor diverso y amalgamado para cada una de ellas. Para poder determinar el número de personas en este grupo poblacional, nosotros utilizamos una pirámide poblacional para la ciudad más poblada del estado, en este caso es Ciudad Juárez con una población estimada de 987,364 habitantes (Estimación para 2012) y determinamos el rango poblacional arriba de 55 años como se muestra en la figura 3, en dicha sociedad son las personas que han arribado de otras sociedades de la Federación, quienes más vulnerabilidad tienen para caer en aislamiento social, esto se debe en gran medida a que no cuentan con familiares directos en la Ciudad, este grupo poblacional compuesto en 2012 por 115,997 habitantes pasará del 14% en 2012 al 27.8% en el 2035 de las personas correspondientes a ese grupo de edad.

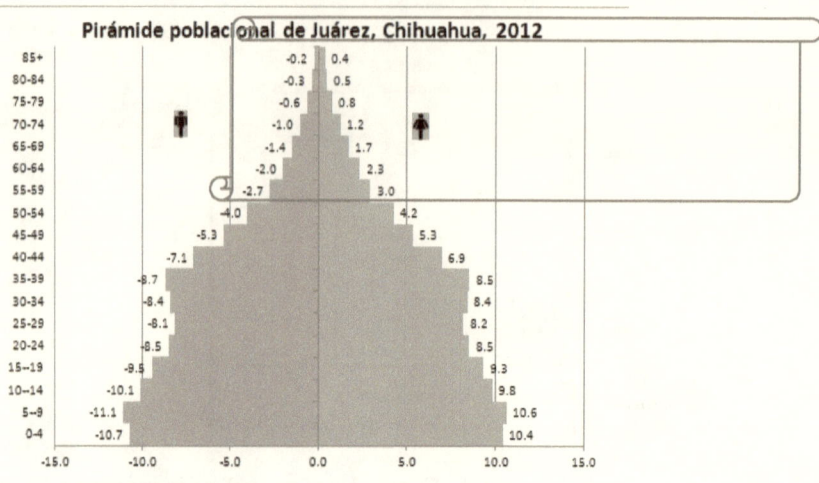

Figura 2.- Pirámide Poblacional en Ciudad Juárez con la distribución de mujeres y hombres con sus respectivos rangos de edad y en un pergamino rojo el grupo de estudio.

Chihuahua tiene políticas públicas para la prevención del aislamiento social, sin embargo en estudios comparados entre diversas fuentes relacionado con la Integración Social, lo que nos ubica en el lugar #27 dentro de la Federación, esto debido a que la falta de cobertura social en áreas rurales del estado, la discriminación y asilamiento social para el grupo de inmigrantes de la Federación, incluyendo a niños y jóvenes de segunda generación, la desnutrición e indicadores de pobreza en grupos indígenas, y aspectos culturales asociados con la religión, incluyendo un imaginario colectivo que la gente que arriba a la ciudad viene buscando únicamente mejorar su situación económica, algo que no es completamente adecuado en una correcta y adecuada comparativa por entidad y edad.

Con respecto a la comparativa con Ciudad Juarez y siguiendo en forma estricta un análisis de Social Data Mining, caracterizamos con datos de la OECD, el índice de Bienestar de Vida para cada una de las 34 Sociedades que conforman dicho Organismo (Ver Figura 3), al analizar la comparativa conformada por 11 atributos: Housing (Acceso a habitación de buena calidad), Income (Ingreso económico), Jobs (Acceso a empleos remunerados basado en los estudios realizados), Community (Organización comunitaria en la Sociedad), Education (Acceso al sistema educativo de cualquier nivel), Environment (Educación Ambiental y cuidado al medio ambiente), Governance (Un Gobierno democrático y al tanto de sus Ciudadanos), Health (Servicios de salud para toda la Sociedad), Life Satisfaction (Satisfacción con las expectativas de vida),

Safety (Seguridad pública ofertada a la Sociedad) y Work-Life Balance (Relación Trabajo-Vida personal en la población económicamente activa) pudimos observar que sólo Turquía presenta indicadores menores a los de México y Ciudad Juárez como Sociedad estaría más cerca de Turquía en lo que respecta a estos atributos sociales, por lo que podemos esperar indicadores similares de aislamiento social, pero al realizar dicho comparativa nos damos cuenta que el Aislamiento Social del grupo de estudio es mayor en 7% para la población mayoritaria y 16% más alta para sus minorías.

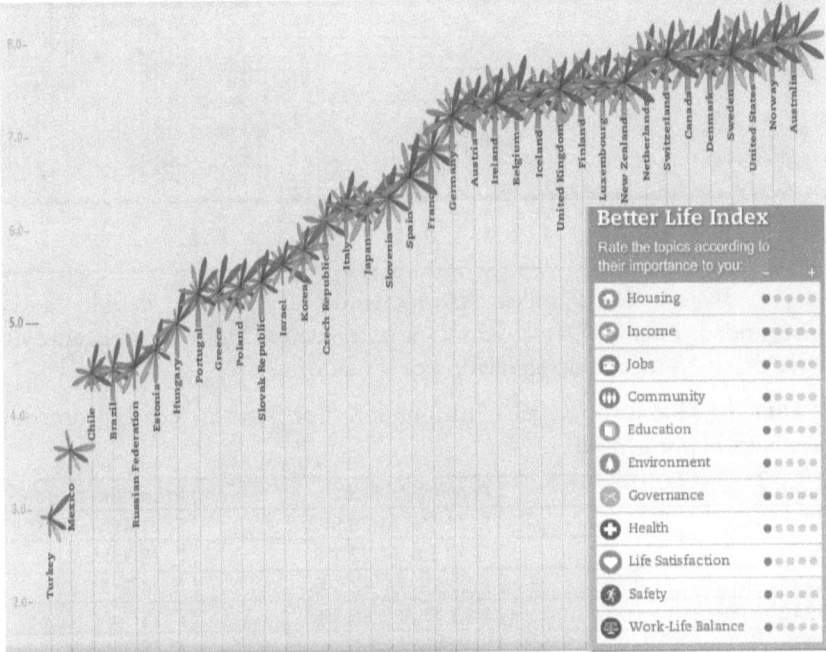

Figura 3.- Representación visual del Indice de Bienestar de Vida para las 34 Sociedades que conforman la OECD.

Otro de los factores que deben ser analizados, es la violencia existente en Ciudad Juárez, la cuál es a su vez la Ciudad más poblada en el estado de Chihuahua, concentrando el 37% de los habitantes del estado, desde 2007 presenta tasas de homicidios arriba de la media nacional y desde 2008 se ha colocado con una de las tasas más altas a nivel internacional teniendo en el 2010 un total de 3116 muertes, y un acumulado de 12,885 muertes para Septiembre de 2012 (Ver Tabla 1), esto ha generado un éxodo masivo que ha afectado el horizonte de planeación de la Ciudad y por ende ha dejado a muchas personas solteras o al menos sin su familia en la Ciudad, debido a que son familias bajo amenaza y con menores posibilidades económicas quienes escapan de la ciudad, un total de 47,415 familias —Esto justificaría en parte la pérdida de al menos 450,000 habitantes-, los cuáles han abandonado la Ciudad como se puede observar en la figura 6.

Familias desplazadas por la Violencia

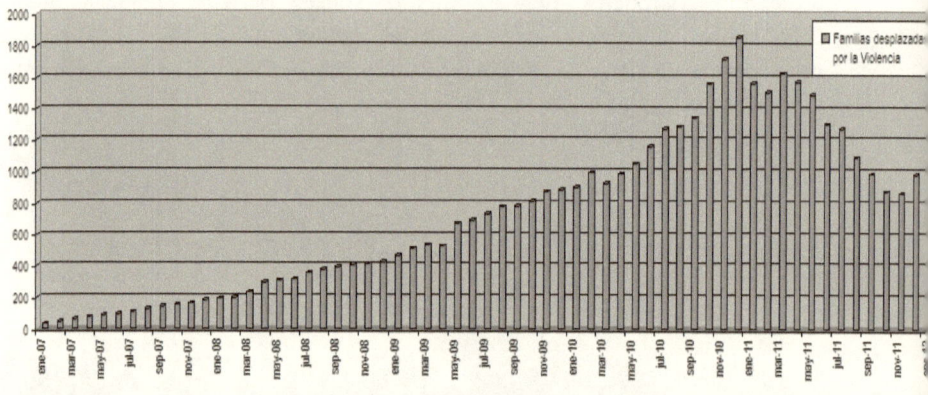

Figura 4.- Familias abandonando la Ciudad debido a la inseguridad. Fuente: Datos del CIS a partir de un Modelo Inteligente de Predicción Numérica desarrollado por los autores.

Tabla 1. Muertes por año, proporción por sexo y edad promedio establecido por género.

Año	Muertes con violencia*	Proporción Sexo	Promedio de edad
2012	1281**	F: 21.5%; M:78.5%*	F: 21.4; M: 20.3*
2011	2047	F: 27.3%; M:72.7%	F: 23.7; M: 21.6
2010	3116	F:16.9%; M: 83.1%	F: 24.2; M: 22.3
2009	2974	F:13.7%; M: 86.3%	F: 25.7; M: 22.6
2008	1783	F:11.4%; M:88.6%	F: 26.2; M: 23.1
2007	615	F:27.2%; M:72.8%	F: 26.7; M: 24.6
Total	12885		

* Fuente Hemerográfica realizada por el CIS y el Observatorio Ciudadano ambos de la UACJ.

** Utilizando un modelo de predicción numérica, ajustado a variables de seguridad y cambio de gobierno a nivel nacional, 717 muertes al 15 de Septiembre del 2012.

Como se puede observar en la Tabla 1, el porcentaje de este tipo de asesinatos es un indicador a la alza desde el 2007 —aunque tuvo una disminución significativa a partir del 2011-, esto sin duda a que ha cambiado radicalmente la edad en la muerte de las víctimas y el cambio en la proporción del sexo de las personas que han sido asesinadas, los datos para el 2012 son el resultado de un Modelo de proyección numérica realizado en el Centro de Investigaciones Sociales de la UACJ, y ajustado con las datos de los años previos durante el presente sexenio de gobierno, aunado a ellos tres rubros importante de violencia continúan a la alza en 2012, el housejacking, el carjacking y el asalto a mano armada en espacios públicos.

Considerando un estudio realizado directamente con el grupo de estudio, tomamos en consideración los que fueron diseñados con varios grupos focales, los cuáles se concentran en la Tabla 2, esta recaba varios aspectos relacionados con la discriminación y el aislamiento social de este grupo social, estas muestras conformadas por 137 individuos (73 mujeres y 64) incluían a adultos hombres y mujeres solteros sin descendencia, con edades comprendidas entre 55 años en adelante: Muestra 1- Gurpo poblacional de 55 a 65 años: 62 (F: 36; M: 26), Muestra 2- Grupo poblacional mayor de 65 años: 55 (F: 25; M: 30), Muestra 3- Grupo de control fuera de Ciudad Juárez: 20 (F: 12; M: 8), y con el fin de convalidar la información, le pedimos a una muestra en otro estado (Morelos) que determinará como percibían el trato hacia personas de Chihuahua en Morelos, si ellos llegarán a vivir con su familia a Morelos y como sería su trato hacia este grupo.

Tabla 2.- Incidencia de actividades familiares relacionadas con un posible Aislamiento Social.

	Muestra1		Muestra 2		Muestra 3	
	F	M	F	M	F	M
N	36	26	25	30	12	8
Asistencia a cumpleaños de familiares directos	58%	55%	45%	31%	85%	74%
Bailar en pareja con algún familiar directo	71%	64%	42%	31%	85%	62%
Participar en la cena navideña	35%	29%	18%	14%	85%	51%
Compartir actividades sociales	72%	89%	69%	51%	80%	70%
Contar historias desde su perspectiva familiar	27%	33%	39%	25%	70%	62%
Decidir la compra de un bien familiar	19%	14%	11%	16%	60%	47%
Incidencia de la Educación de sus familiares	49%	21%	31%	19%	60%	54%
Intercambio lúdico con niños de su familia	59%	17%	22%	10%	70%	51%
Motivación de sus familiares para cumplir metas	73%	47%	56%	34%	70%	43%
Organizar el día de muertos de la familia	68%	43%	57%	28%	85%	50%
Organizar un Convivio Familiar	75%	21%	62%	18%	70%	30%
Organizar una fiesta del Día de la Amistad	65%	40%	43%	13%	75%	22%
Participación de bienes del Capital Social Famiiar	61%	77%	55%	66%	80%	70%
Participar en alguna ceremonia religiosa como padrino/madrina	44%	9%	37%	6%	40%	12%
Participar en un negocio familiar	74%	21%	69%	11%	60%	25%
Participar en un intercambio de regalos en Navidad	93%	73%	82%	68%	85%	62%
Participar en una Pastorela en familia	71%	38%	69%	26%	50%	20%
Preparar una cena navideña	84%	27%	61%	10%	70%	30%
Recibir presentes en sus cumpleaños	97%	72%	81%	56%	90%	70%
Recibir visitas cuando están enfermos	85%	57%	73%	32%	75%	45%

Sentirse apreciados por sus familiares	88%	45%	75%	38%	80%	70%
Ser el responsable de otros familiares de mayor edad	42%	22%	21%	16%	45%	30%
Sugerir actividades para mejorar el capital familiar	81%	77%	89%	70%	80%	70%
Supervisar a otros familiares en una actividad social.	31%	17%	24%	14%	40%	15%
Suma de Reconocimiento Social (Media)	**5.9**	**5.4**	**4.7**	**4..4**	**6.8**	**4.2**

Fuente: Comisión Estatal –Chihuahua- de los Derechos Humanos 2012.

La utilización de Minería de Datos en aspectos sociales ha demostrado ser una tarea clave, para poder corroborar si existen tendencias de aislamiento social para familiares en soltería mayores de 55 años por parte del grupo familiar bajo una situación de abandono social en común, nosotros encontramos variaciones dependiendo del uso de acciones de bloqueo social, aislamiento social y violencia e intimidación (Un tipo de Mobbing Familiar –caracterizado por chantajes emocionales y extorsiones de todo tipo-), ver Tabla 3, cómo se puede apreciar ninguna de las muestras obtiene una media mayor a 6 en la escala de 1 a 10, aunque son las Tías más jóvenes las que alcanzan a ser más apreciados por la mayoría de su familia, considerando que son quienes más han tratado de ser tolerantes con los demás y quienes han apoyado económicamente y moralmente más a sus familias.

En Ciudad Juárez a diferencia con algunas sociedad del resto de la Federación no existe una política pública encaminada a combatir el abandono social derivado de problemas por disimilitud religiosa, económica, política y de espacio físico, la situación que se presenta en Ciudad Juárez no puede justificar como en menos de cinco años la cantidad de abandono social supera por mucho a la media nacional, y con respecto al grupo mayoritario es cercano al 5.7 a 1, lo que en otra Sociedad implicaría una intervención directa por parte de las autoridades gubernamentales para incidir en establecer un Modelo de Convivencia entre las partes, en esta Sociedad no puede ocurrir debido a situaciones legales relacionadas con una Sociedad Multicultural.

3.- Análisis de la muestra analizada relacionada con el abandono social.

Considerando cada uno de los aspectos relevantes asociados con las minorías analizadas desarrollamos la siguiente ecuación (Ecuación 1), la cual pretende justificar las causas del abandono social, esta es definida de la siguiente forma:

$$\text{Índice de Abandono Social} = (\text{Integración Social-Aislamiento Social}) * \sum^{n} (\text{Aspectos Sociales Diversos}) \quad (1)$$

Con base en dicha ecuación, podemos determinar, la ubicación geoespacial -es decir la relación de cada individuo de este grupo social con respecto a los demás- del resto del grupo y determinar las posibilidades de medir el aislamiento social en el tiempo, al visualizar la gráfica (ver Figura 7) podemos comprender que el grupo de mayor vulnerabilidad son los adultos hombres mayores a 65 años, quienes presentan las situaciones económicas más adversas y por ende su aislamiento social es mayor, en el caso de las mujeres mayores de 65 años este varía de acuerdo al tipo de familia en la cual están insertos, esto determina incluso el grado de bloqueo social por parte de sus familiares, por su parte en las muestras de hombres vemos un patrón de apoyo mutuo cuando tienen que convivir con sus respectivas familias, tratan de organizarse para estar juntos y sobrellevar el día a día.

En el caso del grupo analizado en Morelos se presentan aspectos de mayor integración social esto debido a la forma en cómo funcionan las familias extendidas en otras sociedades y el factor religión aumenta la integración social (Ver Figura 5).

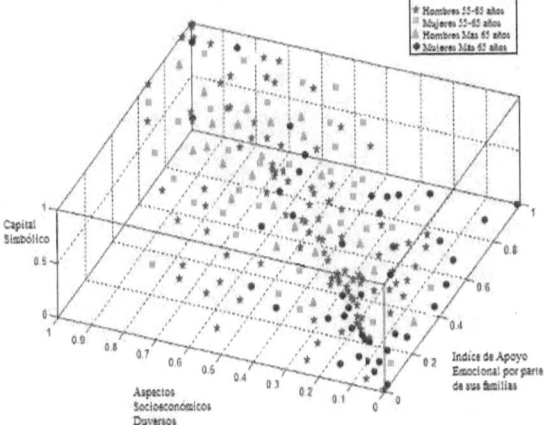

Figura 7.- Representación visual de la muestra analizada, caracterizando el bloqueo social.

Existen diversas situaciones que provocan que acontezca el abandono social, esto debido a la cohesión social y el tiempo transcurrido, en la mayoría de las familias que conforman Ciudad Juárez, no se puede justificar como una sola causa el no contar con hijos e hijas, sino más bien es con respecto a la relación que guarda la convivencia con su familia extendida, entre las diversas causas se encuentran el desempleo en este grupo poblacional que afecta al 32.87% de la población en Chihuahua e incluso en Ciudad Juárez alcanza una cifra de 54.17% para Septiembre del 2012, la otra fundamentalmente esta involucrada con la mala alimentación debido a que en los grupos de mayor pobreza se presentan un consumo calórico menor (37%) al del promedio del grupo mayoritario.

Una vez que fueron entrevistados en forma escrita, la muestra seleccionada se desarrolló un Modelo de Distancia Social de Bogardo (Ver Figura 6), el cual nos reflejó las variantes formas de determinar la situación existente en este grupo poblacional analizada en Ciudad Juárez y como esto afecta a la vida diaria de las personas que no tienen descendencia que cuide de ellos.

Nosotros tomamos en consideración el contexto de los grupos focales, realizados a las muestras minorías, para ello, las agrupamos en tres muestras de acuerdo a su edad y origen para realizar dicha comparativa de los entrevistados utilizando las conversaciones realizadas por la red social caracterizada conformada por cada uno de ellos, esto con la finalidad de identificar diferentes comportamientos sociales (Ver Tabla 3).

Tabla 3.- Distribución de las demandas por categoría y distribuidas para las cinco muestras analizadas.

Categoría	Muestra 1		Muestra 2		Muestra 3	
	F	M	F	M	F	M
N	36	26	25	30	12	8
Imperativas	5%	29%	5%	29%	5%	29%
Directivas Declarativas	5%	6%	5%	6%	5%	6%
Directivas de Simulación	4%	4%	4%	4%	4%	4%
Directivas Interrogativas	2%	7%	2%	7%	2%	7%
Interrogativas de Contexto	35%	16%	35%	16%	35%	16%
Directivas de Conjunción	15%	3%	15%	3%	15%	3%
Cuestionamientos Explosivos	9%	11%	9%	11%	9%	11%
Cuestionamientos Informativos	16%	20%	16%	20%	16%	20%
Mecanismos de atracción de la atención	9%	4%	9%	4%	9%	4%
Total	100%	100%	100%	100%	100%	100%

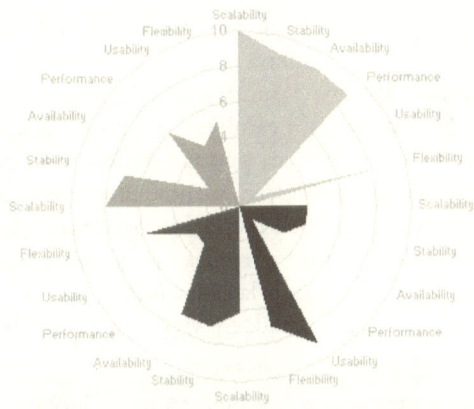

Figura 8.- Gráfico de Radar, mostrando las distancias sociales de Bogardo para los cuatro grupos analizados basados en los datos obtenidos en la Tabla 4.

En la Figura 8, se muestra un estudio comparativo para medir la Distancia Social de Bogardo (Scalability; Stability; Availability; Performance, Usability & Flexibility) utilizando para ello un gráfico de radar midiendo la cohesión social de los rangos de edad que conforman el grupo social analizado, la muestra 1 –Mujeres de 55 a 65 años (color amarillo), la muestra 2 –Hombres de 55 a 65 años (color azul), la muestra 3 (color índigo) –Mujeres mayores de 65 años, y finalmente la muestra 4 (color verde) para hombres mayores a 65 años, y el porcentaje de factor afirmativo a un rubro determinado asociado con el sentimiento de no inclusión y de aislamiento social para poder comprender la perspectiva de ser parte de una familia y como ello afecta el desenvolvimiento emocional.

4.- Consideraciones Finales.

Sentirse útil es importante en todas las etapas de la vida, para un adulto mayor sin descendientes lo suele ser más, es una situación dramática que la mayoría de ellos sólo son buscado bajo situaciones de pedir apoyo económico o algún tipo de ayuda que ellos pueden proporcionar, esta experiencia resulta ser a la larga traumática y difícil de sobrellevar, es por ello que cuando ocurre, las instancias legales y especialmente sociales únicamente tratan de remediar mediante apoyos en especie la situación de este grupo vulnerable. Una política pública incluyente, permitiría ayudar a los grupos sociales vulnerables y poder tener la capacidad de sostener un cambio paradigmático en la sociedad, la cuál debería de ver la edad de sus familiares en primer grado como un vínculo afectivo al formar parte de un "baluarte de capital simbólico familiar".

Se han realizado estudios que indican que las expectativas de este grupo social en cuestión no les permiten acceder a mejores situaciones, lo que conlleva a menores oportunidades en el futuro, lo que finalmente provocará un círculo vicioso en las familias. En cambio, las familias tienen una estructura social arbórea, ramificada, que les permite interactuar con varios tipos de personas en varias categorías en planos diferentes de tiempo y espacio simultáneamente, es por ello que las oportunidades difieren diametralmente acorde a las acciones a realizar para superar el abandono social, un aspecto importante es que la situación de salud es peor que en las personas con familias estructuradas y viven en promedio 10 años menos que el resto de sus familiares.

Debido a las complicaciones existentes en este ambiente de alta violencia que acontece en Chihuahua desde 2007, se ha generado en gran medida, que las personas de este grupo social decidan en forma súbita abandonar el Estado y regresar con sus familias, aunado a que la dinámica social para el resto del Estado haya sufrido cambios drásticos en el aspecto social, es por ello que se debe de buscar una política pública acorde a los cambios poblacionales existentes, debido a que los roles sociales han cambiado y este segmento de la población (Grupo de Estudio) y sus futuras generaciones no podrán ver cumplidos sus sueños e ilusiones de expectativa de vida en esta Sociedad Multicultural y cada vez con un menor número de niños que cuide de ellos.

Referencias

Adderley, R., & Musgrove, P.B. (2001). Data mining case study: Modeling the behavior of offenders who commit serious sexual assaults. *In proceedings of KDD '01*, San Francisco, CA.

Balkin, J. (2006). Law and liberty in virtual worlds. In Balkin, J & Noveck, B (eds.) State of play: Law, Games and virtual worlds. New York: New York University Press.

Chen, H., Chung, W., Xu, J.J., Qin, G.W.Y., & Chau, M. (2004). **Crime data mining: a general framework and some examples**. *Computer*, 17 (4), 50–56.

Garcia-Ruiz, M.A., Vargas Martin, M., Ibrahim, A., Edwards, A., & Aquino-Santos, R. (2009). Combating child exploitation in Second Life. *2009 IEEE Toronto International Conference* – Science and Technology for Humanity (TIC-STH).

Kerbs, R.W. (2005). Social and ethical considerations in virtual worlds. *The Electronic Library*, 23, 539–546.

Schrobsdorff, S. (2007). Predator's playground. *Newsweek*. Available June 14, 2010, from: http://www.newsweek.com/2006/01/26/predator-s-playground.html

Identity and Search in Social Networks

Alberto Ochoa[1,*], Chlöe Malépart[2], Sandra Bustillos[1], and Nemesio Castillo[1]

[1] Juarez City University, México

[2] Université Quebecoise au Montréal, Québec

alberto.ochoa@uacj.mx

Abstract

Social networks have the surprising property of being "searchable": Ordinary people are capable of directing messages through their network of acquaintances to reach a specific but distant target person in only a few steps. We present a model that offers an explanation of social network searchability in terms of recognizable personal identities: sets of characteristics measured along a number of social dimensions. Our model defines a class of searchable networks and a method for searching them that may be applicable to many network search problems, including the location of data files in peer-to-peer networks, pages on the World Wide Web, and information in distributed databases.

Introduction

In the late 1960s, Travers and Milgram (1) conducted an experiment in which randomly selected individuals in Boston, Massachusetts, and Omaha, Nebraska, were asked to direct letters to a target person in Boston, each forwarding his or her letter to a single acquaintance whom they judged to be closer than themselves to the target. Subsequent recipients did the same. The average length of the resulting acquaintance chains for the letters that eventually reached the target (roughly 20%) was about six. This reveals not only that short paths exist (2, 3) between individuals in a large social network but that ordinary people can find these short paths (4). This is not a trivial statement, because people rarely have more than local knowledge about the network. People know who their friends are. They may also know who some of their friends' friends are. But no one knows the identities of the entire chain of individuals between themselves and an arbitrary target.

The property of being able to find a target quickly, which we call searchability, has been shown to exist in certain specific classes of networks that either possess a certain fraction of hubs [highly connected nodes which, once reached, can distribute messages to all parts of the network (5-7)] or are built upon an underlying geometric lattice that acts as a proxy for "social space" (4). Neither of these network types, however, is a satisfactory model of society.

Here, we present a model for a social network that is based upon plausible social structures and offers an explanation for the phenomenon of searchability. Our model follows naturally from six contentions about social networks.

1) Individuals in social networks are endowed not only with network ties, but identities (8): sets of characteristics attributed to them by themselves and others by virtue of their association with, and participation in, social groups (9, 10). The term "group" refers to any collection of individuals with which some well-defined set of social characteristics is associated.

2) Individuals break down, or partition, the world hierarchically into a series of layers, where the top layer accounts for the entire world and each successively deeper layer represents a cognitive division into a greater number of increasingly specific groups. In principle, this process of distinction by division can be pursued all the way down to the level of individuals, at which point each person is uniquely associated with his or her own group. For purposes of identification, however, people do not typically do this, instead terminating the process at the level where the corresponding group size g becomes cognitively manageable. Academic departments, for example, are sometimes small enough to function as a single group but tend to split into specialized subgroups as they grow larger. A reasonable upper bound on group size (9) is $g \cong 100$, a number that we incorporate into our model (Figure 1A). We define the similarity x_{ij} between individuals i and j as the height of their lowest common ancestor level in the resulting hierarchy, setting $x_{ij} = 1$ if i and j belong to the same group. The hierarchy is fully characterized by depth l and constant branching ratio b. The hierarchy is a purely cognitive construct for measuring social distance, and not an actual network. The real network of social connections is constructed as follows.

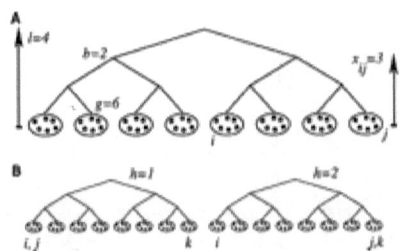

Figure 1

(A) Individuals (dots) belong to groups (ellipses) that in turn belong to groups of groups, and so on, giving rise to a hierarchical categorization scheme. In this example, groups are composed of $g = 6$ individuals and the hierarchy has $l = 4$ levels with a branching ratio of $b = 2$. Individuals in the same group are considered to be a distance $x = 1$ apart, and the maximum separation of two individuals is $x = l$. The individuals i and j belong to a category two levels above that of their respective groups, and the distance between them is $x_{ij} = 3$. Individuals each have z friends in the model and are more likely to be connected with each other the closer their groups are. (**B**) The complete model has many hierarchies indexed by $h = 1 \ldots H$, and the combined social distance y_{ij} between nodes i and j is taken to be the minimum ultrametric distance over all hierarchies $y_{ij} = \min_h x_{ij}^h$. The simple example shown here for $H = 2$ demonstrates that social distance can violate the triangle inequality: $y_{ij} = 1$ because i and j belong to the same group under the first hierarchy and similarly $y_{jk} = 1$ but i and k remain distant in both hierarchies, giving $y_{ik} = 4 > y_{ij} + y_{jk} = 2$.

3) Group membership, in addition to defining individual identity, is a primary basis for social interaction (10, 11) and therefore acquaintanceship. As such, the probability of acquaintance between individuals i and j decreases with decreasing similarity of the groups to which they respectively belong. We model this by choosing an individual i at random and a link distance x with probability $p(x) = c \exp[-\alpha x]$, where α is a tunable parameter and c is a normalizing constant. We then choose a second node j uniformly among all nodes that are a distance x from i, repeating this process until we have constructed a network in which individuals have an average number of friends z. The parameter α is therefore a measure of homophily—the tendency of like to associate with like. When $e^{-\alpha} \ll 1$, all links will be as short as possible, and individuals will connect only to those most similar to themselves (i.e., members of their own bottom-level group), yielding a completely homophilous world of isolated cliques. By contrast, when $e^{-\alpha} = b$, any individual is equally likely to interact with any other, yielding a uniform random graph (12) in which the notion of individual similarity or dissimilarity has become irrelevant.

4) Individuals hierarchically partition the social world in more than one way (for example, by geography and by occupation). We assume that these categories are independent, in the sense that proximity in one does not imply proximity in another. For example, two people may live in the same town but not share the same profession. In our model, we represent each such social dimension by an independently partitioned hierarchy. A node's identity is then defined as an H-dimensional coordinate vector $\vec{v}_i$, where v_i^h is the position of node i in the hth hierarchy, or dimension. Each node i is randomly assigned a coordinate in each of H dimensions and is then allocated neighbors (friends) as described above, where now it

randomly chooses a dimension h (e.g., occupation) to use for each tie. When $H = 1$ and $e^{-\alpha} \ll 1$, the density of network ties must obey the constraint $z < g$.

5) On the basis of their perceived similarity with other nodes, individuals construct a measure of "social distance" y_{ij}, which we define as the minimum ultrametric distance over all dimensions between two nodes i and j; i.e., $y_{ij} = \min h \times_{ij} h$. This minimum metric captures the intuitive notion that closeness in only a single dimension is sufficient to connote affiliation (for example, geographically and ethnically distant researchers who collaborate on the same project). A consequence of this minimal metric, depicted in Figure 1B, is that social distance violates the triangle inequality—hence it is not a true metric distance—because individuals i and j can be close in dimension h_1, and individuals j and k can be close in dimension h_2, yet i and k can be far apart in both dimensions.

6) Individuals forward a message to a single neighbor given only local information about the network. Here, we suppose that each node i knows only its own coordinate vector $\vec{v}_i$, the coordinate vectors $\vec{v}_j$ of its immediate network neighbors, and the coordinate vector of a given target individual $\vec{v}_t$, but is otherwise ignorant of the identities or network ties of nodes beyond its immediate circle of acquaintances.

Individuals therefore have two kinds of partial information: social distance, which can be measured globally but which is not a true distance (and hence can yield misleading estimates); and network paths, which generate true distances but which are known only locally. Although neither kind of information alone is sufficient to perform efficient searches, here we show that a simple algorithm that combines knowledge of network ties and social identity can succeed in directing messages efficiently. The algorithm we implement is the same greedy algorithm Milgram suggested: Each member i of a message chain forwards the message to its neighbor j who is closest to the target t in terms of social distance; that is, y_{jt} is minimized over all j in i's network neighborhood.

Our principal objective is to determine the conditions under which the average length $\langle L \rangle$ of a message chain connecting a randomly selected sender s to random target t is small. Although "small" has recently been taken to mean that $\langle L \rangle$ grows slowly with the population size N(13, 14), Travers and Milgram found only that chain lengths were short. Furthermore, these message chains had to be short in an absolute sense because at each step, they were observed to terminate with probability $p \cong$ 0.25 (1, 15). We therefore adopt a more realistic, functional notion of efficient search, defining for a given message failure probability p, a searchable network as any network for which q, the probability of an arbitrary message chain reaching its target, is at least a fixed value r. In terms of chain length, we formally require $q = \langle (1-p)L \rangle \geq r$, and from this we can obtain an estimate of the maximum required $\langle L \rangle$ using the approximated inequality $\langle L \rangle \leq \ln r / \ln(1-p)$. For the purposes of this study, we set $r = 0.05$ and $p = 0.25$, yielding the stringent requirement that $\langle L \rangle \leq 10.4$ independent of the population size N. Figure 2. A presents a typical phase diagram in H and α, outlining the searchable network region for several choices of N, $g = 100$, and $z = g - 1 = 99$.

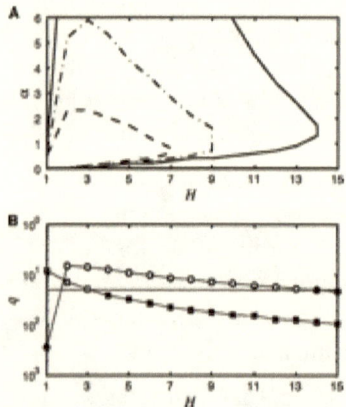

Figure 2

(A) Regions in H-α space where searchable networks exist for varying numbers of individual nodes N (probability of message failure $p = 0.25$, branching ratio $b = 2$, group size $g = 100$, average degree $z = g - 1 = 99$, 10^5 chains sampled per network). The searchability criterion is that the probability of message completion q must be at least $r = 0.05$. The lines correspond to boundaries of the searchable network region for $N = 102{,}400$ (solid), $N = 204{,}800$ (dot-dash), and $N = 409{,}600$ (dash). The region of searchable networks shrinks with N, vanishing at a finite value of N that depends on the model parameters. Note that $z < g$ is required to explore H-α space because for $H = 1$ and α sufficiently large, an individual's neighbors must all be contained within their sole local group. (B) Probability of message completion $q(H)$ when $\alpha = 0$ (squares) and $\alpha = 2$ (circles) for the $N = 102{,}400$ data set used in (A). The horizontal line shows the position of the threshold $r = 0.05$. Open symbols indicate that the network is searchable ($q \geq r$) and closed symbols mean otherwise. For $\alpha = 0$, searchability degrades with each additional hierarchy. For the homophilous case of $\alpha = 2$ with a single hierarchy, less than 1% of all searches find their target ($q \cong 0.004$). Adding just one other hierarchy increases the success rate to $q \cong 0.144$, and q slowly decreases with H thereafter.

Our main result is that searchable networks occupy a broad region of parameter space (α, H) which, as we argue below, corresponds to choices of the model parameters that are the most sociologically plausible. Hence our model suggests that searchability is a generic property of real-world social networks. We support this claim with some further observations and demonstrate that our model can account for Milgram's experimental findings.

First, we observe that almost all searchable networks display $\alpha > 0$ and $H > 1$, consistent with the notion that individuals are essentially homophilous (that is, they associate preferentially with like individuals) but judge similarity along more than one social dimension. Neither the precise degree to which they are homophilous, nor the exact number of dimensions they choose to use, appears to be important—almost any reasonable choice will do. The best performance, over the largest interval of α, is achieved for $H = 2$ or 3—an interesting result in light of empirical evidence (16) that individuals across different cultures in small-world experiments typically use two or three dimensions when forwarding a message.

Second, as Figure 2B shows, although increasing the number of independent dimensions from $H = 1$ yields a dramatic reduction in delivery time for values of $\alpha > 0$, this improvement is gradually lost as H is increased further. Hence the window of searchable networks in Figure 2A exhibits an upper boundary in H. Because ties associated with any one dimension are allocated independently with respect to ties in any other dimension, and because for fixed average degree z, larger H necessarily implies fewer ties per dimension, the network ties become less correlated as H increases. In the limit of large H, the network becomes essentially a random graph (regardless of α) and the search algorithm becomes a random walk. An effective decentralized search therefore requires a balance (albeit a highly forgiving one) of categorical flexibility and constraint.

Finally, by introducing parameter choices that are consistent with Milgram's experiment ($N = 10^8, p = 0.25$) (1), as well as with subsequent empirical findings ($z = 300, H = 2$) (16, 17), we can compare the distribution of chain lengths in our model with that of Travers and Milgram (1) for plausible values of α and b. As Figure 3 shows, we obtain $\langle L \rangle \cong 6.7$ for $\alpha = 1$ and $b = 10$, indicating that our model captures the essence of the real small-world problem. This agreement is robust with respect to variations in the branching ratio, showing little change over the range $5 < b < 50$.

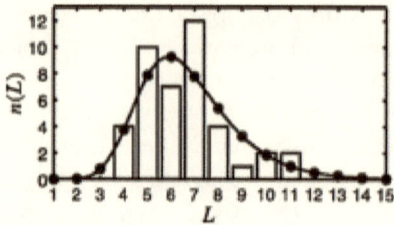

Figure 3

Comparison between $n(L)$, the number of completed chains of length L, taken from the original small-world experiment (1) (bar graph) and from an example of our model with $N = 10^8$ individuals (filled circles with the line being a guide for the eye). The experimental data shown are for the 42 completed chains that originated in Nebraska. (We have excluded the 24 completed chains that originated in Boston as this would correspond to $N \cong 10^6$.) The model parameters are $H = 2$, $\alpha = 1$, $b = 10$, $g = 100$, and $z = 300$; message attrition rate is set at 25%; $n(L)$ for the model is compiled from 10^6 random chains and is normalized to match the 42 completed chains that started in Nebraska. The average chain length of Milgram's experiment is ~6.5, whereas the model yields $\langle L \rangle \cong 6.7$. The distributions compare well: A two-sided Kolmogorov-Smirnov test yields a P-value of $P \cong 0.57$, whereas for a χ^2 test, $\chi^2 \cong 5.46$ and $P \cong 0.49$ (seven bins). (A large value of P supports the hypothesis that the distributions are similar.) Even without attrition, the model's average search time is $\langle L \rangle \cong 8.5$ and the median chain length is 8. The model does not entirely match the experimental data because the former requires approximately 360 initial chains to achieve 42 completions as compared with 196.

Although sociological in origin, our model is relevant to a broad class of decentralized search problems, such as peer-to-peer networking, in which centralized servers are excluded either by design or by necessity, and where broadcast-type searches (i.e., forwarding messages to all neighbors rather than just one) are ruled out because of congestion constraints (6). In essence, our model applies to any data structure in which data elements exhibit quantifiable characteristics analogous to our notion of identity, and similarity between two elements—whether people, music files, Web pages, or research reports—can be judged along more than one dimension. One of the principal difficulties with designing robust databases (18) is the absence of a unique classification scheme that all users of the database can apply consistently to place and locate files. Two musical songs, for example, can be similar because they belong to the same genre or because they were created in the same year. Our model transforms this difficulty into an asset, allowing all such classification schemes to exist simultaneously, and connecting data elements preferentially to similar elements in multiple dimensions. Efficient decentralized searches can then be conducted by means of simple, greedy algorithms providing only that the characteristics of the target element and the current element's immediate neighbors are known.

REFERENCES

J. Travers, S. Milgram, Sociometry **32**, 425 (1969).

D. J. Watts, S. J. Strogatz, Nature **393**, 440 (1998).

S. H. Strogatz, Nature **410**, 268 (2001).

J. Kleinberg, Nature **406**, 845 (2000).

A.-L. Barabási, R. Albert, Science **286**, 509 (1999).

L. Adamic, R. Lukose, A. Puniyani, B. Huberman, Phys. Rev. E **64**, 046135 (2001).

B. J. Kim, C. N. Yoon, S. K. Han, H. Jeong, Phys. Rev. E **65**, 027103 (2002).

H. C. White, **Identity and Control** (Princeton Univ. Press, Princeton, NJ, 1992).

G. Simmel, Am. J. Sociol. **8**, 1 (1902).

F. S. Nadel, **Theory of Social Structure** (Free Press, Glencoe, IL, 1957).

R. Breiger, Social Forces **53**, 181 (1974).

B. Bollobás, **Random Graphs**(Academic Press, New York, 1985).

M. Newman, D. Watts, Phys. Rev. E 60, 7332 (1999).

J. Kleinberg, **Proc. 32nd ACM Symposium on Theory of Computing** (Association for Computing Machinery, New York, 2000).

H. C. White, Social Forces **49**, 259 (1970).

H. Bernard, P. Killworth, M. Evans, C. McCarty, G. Shelly, Ethnology **27**, 155 (1 988).

P. Killworth, H. Bernard , Soc. Networks **1**, 159 (1978).

B. Manneville, **The Biology of Business: Decoding the Natural Laws of the Enterprise**(Jossey-Bass, San Francisco, 1999), chap. 5.

Evaluating an Asian Film Festival using a strategic Hybrid Algorithm

Alberto Ochoa[1], Daniel Azpeitia[1], Rubén Jaramillo[2] & Chlöé Malépart[3]

[1]Universidad Autónoma de Ciudad Juárez

[2]CFE-LAPEM, México

[3] Université Quebécoise au Montreal.

alberto.ochoa@uacj.mx

Abstract

In this research analyze the voting behavior in a specialized Film Festival held every year in San Francisco. The use of Data Mining and Bioinspired algorithm permits to analyze the determinants of success from a specific kind of film. We show that they are rather driven by social and cultural proximities between Asian Films and people in Western Societies. With this information is possible to predict the popularity of a specific film, calculated the assigned votes from the jury, this paper tries to explain this social behavior in an Asian Film Festival.

Keywords: Particle Swarm Optimization, Social Data Mining and Social Modelling.

1. Introduction

International Asian American Film Festival held for the first time in San Francisco, in 1982 with seven films from Japan and South Korea, now films from Israel, Turkey, Armenia, Georgia, and Azerbaijan are now regular participants, inclusively Palestine, Tíbet, Abkhazia and seven Russian Republics had take participation in the Festival . Since 2002 exist eleven different categories. Each film is promoted by different media; since 2001 the contest of films related with the festival is broadcast live to the people select the best films in each category. Nowadays, it is watched by several hundred of people. The ratings are normalized in a range of 1 to 7 using a Lickert Scale. This allows each voting people to give positive ratings to seven films. The order in which the films are showed is randomly drawn before the competition starts. After the films are showed, people are asked to cast their votes. Results are announced one day later to calculate the proportion of votes. The films are ranked according to their aggregate score. Many contests competing by a best place as Eurovision have been studied with different perspectives: the compatibility between countries and the

political and cultural structures of Europe [4], the persistent structure of hegemony in the Eurovision Song Contest [5], cultural voting [6] and the analysis about Grand Prix which evaluate many countries participating in different years and with different many of countries competing [7], among others. This research is novel because analyze the behavior of people when films from different cultures participate in this Festival. The objective is will estimate the final ranking of these films. The organization of this article is the following. The analysis of the 30 Asian Film Festivals editions to incorporate *a priori* knowledge about the voting patterns and relationships between potential winners is explained in Section 2. Next, the problem statement is defined in Section 3. The COPSO algorithm is thoroughly explained in Section 4. In Section 5, our approach is tested in the new Asian Film Festival 2012. The experiments and the analysis applied to estimate the ranking of a specific film in the new Festival are explained in Section 6. Conclusions are provided in Section 7.

2. Asian Film Festival Ranking using Data Mining

Data mining is the search of global patterns and the existent relationships among the data of immense databases, but that are hidden in them inside the vast quantity of information [3]. These relationships represent knowledge of value about the objects that are in the database. This information is not necessarily a faithful copy of the information stored in the databases. Rather, is the information that one can deduce from the database. One of the main problems in data mining is that the number of possible extracted relationships is exponential [2]. Therefore, there are a great variety of machine's learning heuristics that have been proposed for the discovery of knowledge in databases [2]. One of the most popular approaches to represent the results of data mining is to use decision trees. A decision tree provides a procedure to recognize a given case for a concept. It is a "divided and conquer" strategy for the acquisition of the concept (instance). The decision trees have been useful in a great variety of practical cases in science and engineering; in our case we use data mining to characterize the historical voting behavior for each country. Thus, we selected societies that have participated and characterized its behavior based on its votes previously emitted, which allowed to describe so much to the society as to the individual. The purpose is to explain v_{ij}, the vote (that is, the number of points) cast by the people of country $i \in L$ in evaluating the performer of a film $j \in L$ ($i \neq j$, since a person only can vote for a specific film), where L is the total number of films. Without taking into account any other feature, the voting equation could simply be written

$$v_{ji} = \alpha_{ij}v_{ij} + u_{ij} \qquad (1)$$

Where α_{ij} is a commitment parameter, and v_{ij} a random disturbance. If exchanges of votes were "perfect", and both films kept their commitment, α_{ij} would be equal to 1. More generally, such an equation should contain variables $k=\{1,\ldots,K\}$ representing the characteristics (language in which is presented the film, thematic, sound track, photography and others) of a film i, and variables representing the performances of this i along its Ti participation in the Asian Festival.

$$v_{ji} = \alpha_{ij}v_{ij} + \beta\sum_{k=1}^{K} x_{ik} + \gamma\sum_{t=1}^{T_i} z_{it} + u_{ij} \qquad (2)$$

where β and γ are parameters to be estimated. The part associated with beta parameter is related with the attributes of performance of a film. The part associated with gamma parameter is related with the performance of these films during Asian Festival's participations. A problem is concerned with the fact that will appear on the other side of the equation for the observation concerning the vote of a person i for the film representing a specific country j. This can be dealt with in several ways. First, and this is the easiest way, instead of using v_{ij} in the right-hand side, one can use the vote cast in previous competition, say v_{ij}^{-1}, though one could think that countries would not necessarily keep their commitment over time. An alternative is to use only half of the observations along all Asian Festivals editions; thus, very v_{ij} that appears in the right-hand side of the equation is not used in the left-hand side. The voting equation is estimated by linear methods. The influence of the order in which sound track appear in competition has often been outlined. The exogenous order in which films perform is thus included as determinant. Other variables include (a) a dummy for new films, this variable takes the value 1 for the performer whose films are the same ancestors that of the host country-, (b) the language, in which the films is make, (c) thematic, and (d) whether the film using a specialized photography. The last group of variables will include linguistic and cultural distances between voters and films, and may dispense us from using variables that characterize voters. National culture differences are represented by the four dimensions studied in [1]. These studies identified a scored the four following dimensions that make for "cultural distances":

(a)   Power Distance: It measures the extent to which the less powerful members of a society accept that power is distributed unequally; it focuses in the degree of equality between individuals;

(b)   Individualism: It measures the degree to which individuals in a society are integrated into group; it focuses on the degree a society reinforces individual or collective achievement and interpersonal relationships;

(c)   Masculinity: It refers to the distribution of roles between genders in a society; it focuses on the degree a society reinforces the traditional masculine work role of male achievement, control, and power;

(d)   Uncertainty Avoidance: It deals with a society's tolerance for uncertainty or ambiguity, and refers to man's search for truth.

Table 2: Cultural Distances vs Contender Characteristics

	(a)	(b)	(c)	(d)
Quality	0.911	0.914	0.901	0.905
	(0.03)	(0.03)	(0.03)	(0.03)
Logrolling	0.028	0.022	0.018	0.016
	(0.01)	(0.01)	(0.01)	(0.01)
Order of perf.	0.003	0.002	0.004	0.003
	(0.01)	(0.01)	(0.01)	(0.01)
Host country	0.177	0.191	0.155	0.171
	(0.24)	(0.24)	(0.24)	(0.24)
Sung in english	0.14	0.193	0.101	0.135
	(0.14)	(0.14)	(0.14)	(0.14)
Sung in french	0.353	0.354	0.343	0.347
	(0.17)	(0.17)	(0.18)	(0.18)
Male singer	0.139	0.148	0.147	0.154
	(0.13)	(0.13)	(0.13)	(0.13)
Duet	0.223	0.147	0.203	0.174
	(0.20)	(0.20)	(0.20)	(0.20)
Group	0.1	0.08	0.087	0.079
	(0.13)	(0.13)	(0.13)	(0.13)
Language	-	-1.142	-	-0.634
		(0.22)		(0.24)

Table 1: Correlations between Cultural Distances and Linguistic

	Language	Power	Indiv.	Masc.	U.A.
Language	1				
Power	0.205	1			
Indiv.	0.254	0.111	1		
Masc.	-0.002	0.031	-0.128	1	
U.A.	0.319	0.367	0.404	0.083	1

$$r_i = 0.4 \sum_{k=1}^{7} x_{ik} + 0.6 \sum_{t=1}^{T_i} z_{it}$$

Table 1 illustrates the correlations between the cultural distance and native languages for the countries that are present in our sample. Uncertainty avoidance is correlated with three other variables, but otherwise, distances seem to pick very different dimensions of people's behavior. The generated configurations can metaphorically be related to the knowledge of the behavior of the community with respect to an optimization problem (to make alliances to obtain a better ranking). Columns (a) to (d) of Table 2 contain the results of an OLS estimation of equation 2.

We first observe that quality always plays a very significant role, which should of course not be surprising. Logrolling is significant only in (a), in which no account is taken of linguistic and cultural distances. It ceases to be so in all the other equations once linguistic and/or cultural distances are also accounted for. Note that even when the coefficient is significantly different from zero, its value is very small. Order of appearance plays no role, while among the other variables, the only one which has some influence is "this film is showed in Urdu Language". Though not all distance coefficients are significantly different from 0 at the 5 percent probability level, they all pick negative signs (the larger the distance, the lower the rating). The Table 3 presents the expected performance rates for 2009. The performance rate tries to predict the film rank through environment variables observed along 30 Asian Film editions. In 2011 participated 47 societies hence it was more complex to obtain a second place, than in opposition, a film that obtained second place in Asian Film Festival in 1992 when only seven societies participated. Obviously, for all films exists historical information of 30 Film Festival. The information obtained through data mining, denotes a similar behavior of films with similar characteristics (language, territorial extension, religion, in others). Thus, the historical performance for each film was calculated whit itself. The parameters used by the model to calculate the performance rate are: $\beta=0.4$ and $\gamma=0.6$

3. Problem Statement

The objective of this paper is to estimate the position rank of a new film. This implies to estimate the final voting matrix, where every cell j, i represents the score gives to each film i by the people j; that is v_{ji}. For attaining a well prediction, the model should controls the voting behavior between judges and people taking account the historical performance that reflects the cultural empathy, the commonality of regions. The estimated performance rate could guide the model towards an optimal voting configuration according to the current expectations of the experts.

The next objective function posses these two important features of the Asian Film Festival, the voting behavior and the performance rate explained in the previous Section. Notice that Equation 3 is part of Equation 4.

$$f = \sum_{i=1}^{C}\sum_{j=1}^{N} c_{ij} + 4\sum_{i=1}^{C}\sum_{k=1}^{S} p_{ik} + \frac{2}{max_S}\sum_{i=1}^{C} s_i * r_i$$

Maximize   (3)

Subject to:

- Film j cannot vote for itself.

- People j just can vote one time for contender i.

- People j just can give a score k to only one contender i.

Where N is the number of voting people, C is the number of films, S is the number available scores $S=\{7,6,5,4,3,2,1\}$ and $max_s=7$ is the maximum score. The first two terms represents the performance of the final ranking. In the first term of Equation 3, c_{ij} is the probability that a score k was given by people j for a film i. The probability for each film was calculated observing the behavior of the voting along 30 Asian Film Festival editions. The model explained in this section, implies to solve a combinatorial problem which attempts to estimate the final voting. The constrained optimization problem has two parts. In the first part, the problem is to find the optimal combination that maximizes the sum of probabilities (first two terms of Equation 4). This implies the totally of people voting (subject to the mentioned constraints) which must assign 7 different scores (S) to each film, resulting 1.87E+14 possible combinations. In the second part, the total sum of the votes obtained by every film is calculated. The vote sums (s_i) are used to calculate the weighted sum presented in Equation 3 (third term). This implies again to find the optimal combination out of 1.87E+14 possible solutions. The maximization of both parts of the problem generates a tradeoff between the voting behavior and the performance rate. For solving the current optimization problem, we use a simple and innovative PSO for solving constrained optimization problems which is thoroughly explained in the next section.

4. Constrained optimization via PSO

Particle Swarm Optimization (PSO) [2] algorithm is inspired by the motion of a bird flock. A member of the flock is called "particle". In PSO, the source of diversity, called *variation*, comes from two sources. One is the difference between the particle's current position x_t and the global best G_{Best} (best solution found by the flock), and the other is the difference between the particle's current position x_t and its best historical value P_{Best} (best solution found by the particle). Although variation provides diversity, it can only be sustained by for a limited number of generations because convergence of the flock to the best is necessary to refine the solution. The velocity equation combines the local information of the particle with global information of the flock, in the following way.

$$v_{t+1} = w * v_t + \phi_1 * (P_{Best} - x_t) + \phi_2 * (G_{Best} - x_t)$$
$$x_{t+1} = x_t + v_{t+1} \qquad (5)$$

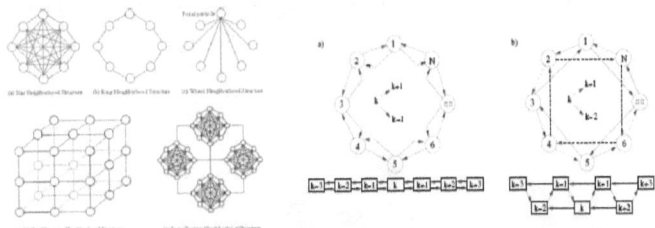

Figure 1: Neighborhood structures for PSO.

Figure 2: Ring neighborhood structures. a)Doubly-linked ring (original PSO). b)Singly-linked ring

A leader can be global to all the flock, or local to a flock's neighborhood. Flock neighborhoods have a structure that defines the way information is concentrated and then distributed among its members. The most common flock organizations are shown in Figure 1. The organization of the flock affects search capacity and convergence. The original ring structure is implemented by a doubly-linked list, as shown in Figure 2-a. COPSO uses an alternative ring implementation, the singly-linked list, shown in Figure 2-b. This structure improved the success of experimental results by a very important factor.

5. Experiments using a Model Validation

For knowing the performance of the proposed model, it was used to estimate the final votes of a specific film named "White Frog" from Macau. This film competed in the category of complete film against other 13 films from different countries. For estimating the voting matrix, 30 runs of the each experiment, Final, were performed to obtain a better estimation of the final ranking. In every run, 350,000 function evaluations were performed. The average along the 30 runs was calculated for every film. Next, the average ranking was obtained to determine the 10 films which best are going to contend in the Final ranking. Three measures are calculated from the 30 runs: average, median, and interquartile range. The interquartile range has a comprehensiveness of 50% around the median value (second quartile Q2), which is calculated through the lower quartile Q1 (first quartile) and upper quartile Q3 (third quartile). In descriptive statistics a quartile is any of the three values which divide the sorted data set into four equal parts, so that each part represents $1/4^{th}$ of the sampled population. The difference between the upper and lower quartiles is called the interquartile range. In section 6, the estimation of our approach for the new Asian Film Festival is presented.

6. Experiments in Asian Film Festival.

In 2012, the Asian Film Festival consists of 11 different categories. The objective of this experiment is to predict the final ranking for each film. For this experiment 30 runs were performed with 27,000 function evaluations. The top-10 of the 30 runs indicates that just the film selected will be in the three with more votes. For estimating the final ranking of this film, 30 runs were performed with 27,000 function evaluations. The average median and interquartile range for the 30 runs was calculated. The experiments predict correctly the final positions.

Figure 4: Picture of "White Frog" from Macau.

7. Conclusions

The prediction of future events is a hard task, also, impossible in several topics [8]. There are a several methods that have been used as an auxiliary tool, for building estimation models. In this work, data mining and evolutionary computation are combined for predicting the behavior in a Film Festival as in [1]. Our approach proposes a model that includes two main features: voting behavior and cultural characteristics [9]. The model incorporates historical information about the vote assignation, that people has performed along previously editions. Besides the model includes information about intrinsic characteristics of the contender that represents each film.

References

[1] V. Ginsburgh and A. Noury. Cultural voting: The Eurovision Song Contest. http://ssrn.com/abstract=884379, 2005.

[2] J. Kennedy and R. Eberhart. The Particle Swarm: Social Adaptation in Information-Processing Systems. McGraw-Hill, London, 1999.

[3] A. Ochoa et al. Italianitá: Discovering a Pygmalion effect on Italian communities using data mining. In Proceedings of CORE'2006.

[4] M. Rauhlen. Culture's Consequences. Beverly Hils, California: Sage, 1997.

[5] T. Suaremi, K. Shikari & Shayera Hal. Understand social groups using artificial intelligence techniques. In Proceedings of NDAM'2006, Reykiavik, Iceland, 2006.

[6] G. Yair. Unite unite Europe: The political and cultural structures of Europe as relected in the eurovision song contest. Social Netwroks, 17(2): 147-161, 1995.

[7] G. Yair and D. Maman. The persistent structure of hegemony in the Eurovision Song Contest. Acta Sociologica, 39:309-325, 1996.

[8] D. Zolezzi, Dori Aandraison & A. Ochoa-Zezzatti. A model to explain the extinction of San Benedicto Rock Wren using Cultural Algorithms. In Proceedings of OCAAI'2007. Bakú, Azerbaijan, 2007.

[9] A. Noudher. Palestine in Eurovision. Master Thesis of Sociology of Islamic University of Gaza, 2007.

Podcast's preferences used on information, entertainment, and socializing.

Alberto Ochoa1,*, Rubén Jaramillo2, Sandra Bustillos3, Daniel Azpeitia1, and Saúl González1

1 Juarez City University, México

2 CFE-LAPEM.

alberto.ochoa@uacj.mx

Abstract

Millions of listeners tune in to internet-based audio-on-demand programming (commonly known as audio podcasts). College students represent a large and influential group of consumers but little is known about this primary audience's podcast listening choices and related options. This research was organized within the uses and gratifications perspective and involved an online survey of college students who identified as podcast listeners. Results showed most collegiate listeners spent no more than an hour a month listening to podcasts, and fewer than half reported listening to all the episodes they downloaded. Most listeners found podcasts fun and entertaining and said they enjoy sharing what they enjoy sharing what they have head with peers. Respondents reported clearly indentified program genre preferences, and reported "multitasking" in a variety of specific ways while listening. This research does not attempt to answer all questions about the college student podcast audience, but it is a starting point for further study into this important user group. Recommendations for future research are offered.

1. Introduction

The creation and use of digital on-demand audio exploded into popular culture in the fall of 2001, when Apple introduced the iPod digital audio player. The device allowed computer users to download audio files from the Internet and then listen to the files at their leisure. These files, which allowed broadcasts to be accessed and used at the discretion of the listener, became known as "podcasts" (Notess, 2005).

In the short span of time since the iPod's introduction, the Apple player and a host of imitators that also play podcast audio have gone from novelty items to must have technology for millions. It is rare to venture into any public place today and not see someone listening to a digital audio player.

Undoubtedly, part of the reason for the ubiquity of the digital audio player is that podcast technology is such a good fit with contemporary consumer culture. Podcast listeners are empowered listeners. They have complete freedom to review, select, download, and listen to programming on a myriad of subjects any time they wish. Much of the programming is available at no cost.

One researcher estimated that the 24/7 marketplace of the Internet is home to at least 30,000 podcast programs (Chung, 2008). The number of programs and episodes appears to be growing rapidly. Apple's iTunes Store podcast locator was recently expanded to 21 categories. The iTunes locator lists hundreds of podcast programs, some with dozen of episodes available (Tips for podcast, 2009).

Podcast technology seems especially attractive to a large and growing population of college students. There were an estimated 18.2 million college students in the U.S. alone in 2007, an increase of 26% since 1997 (U.S. Department of Education, 2009). Although student demographics are shifting, the vast majority of college students are fulltime undergraduates between 18 and 24 years of age. These students comprise an overwhelmingly young, mobile, affluent and technologically savvy public.

An Alloy Media+Marketing, 675 survey of college student buying power showed college students to be "an empowered group of consumers" (College students setting, 2008) who would spend $237 billion on consumer goods in 2008. Among college students surveyed by Alloy Media+Marketing, 67% reported ownership of a digital audio player. Another marker research survey of college students concluded that the iPod music player "surpassed beer drinking as the most 'in' thing among undergraduate college students" (Survey: Apple iPod, 2006) One researcher went so far as to declare that college students "maybe the main user group of the podcast" (Chung, 2008).

Despite the tremendous popularity of digital audio players and content on college campuses and elsewhere, "almost nothing is known about listeners, their habits, or podcast demand" (Brown, 2006). A large collection of general interest and scholarly literature confirms that audio podcasts are easy to create, can help market products and deliver news. A growing body of work focuses on podcasts' potential for bolstering educational curricula and increasing the potential for student learning. A similarity large body of collection of research addresses the technical issues relating to podcast content. But there is almost no research literature that identifies the type of people most likely to listen to podcasts, the types of programming people are listening to, and the ways listeners incorporate podcast content into their lives (Chung, 2008).

Because the need for insight into podcast use by college students is particularly noteworthy, the present study was undertaken to generate some initial findings about how college students would report incorporating the use of digital audio into their lives. The stud was conducted within the framework of uses and gratification theory (Couche, 1996; Fisher, 1978). The study gathered information about active audience members who interact with a medium in a different ways. These audience members use media content to fulfill a variety of perceived needs (Ruggiero, 2000).

This research does not attempt to address all the relevant questions about collegiate listeners or the programming they are listening to. It does, however, provide some valuable understanding about student's digital audio download habits, program preferences, and actions taken to share with others The study allows for some insight into listeners' perceptions of usefulness and entertainment value of podcast programming, an area neglected in the literature.

2.   Literature review

People use the Internet to seek information, meet others, pass time, be entertained, and add convenience to their lives (Paparachissi & Rubin, 2000). Although the technology to transmit audio content over the Internet is not new, the explosion in on-demand audio activity is a recent phenomenon that can be traced to Apple's introduction of the iPod digital audio player in the fall of 2001. Though the iPod is not the only device capable of playing mp3 format audio, a 2006estimate found more than 70% of digital audio players sold were iPods (Marsal, 2006) and by 2007 more than 100 million iPods had been sold (100 million, 2007).

As the world's consumers snapped up iPod players, a dramatic escalation in the creation of podcast audio content took place. The totality of available content has increased so rapidly that it is difficult to get accurate indications of the number of podcast programs (series) and episodes (individual broadcasts within a series).

There were only a few thousands podcast episodes available on the Internet in January of 2005. Within the year the number had grown to more than 25,000 (Bullis 2005) and by 2008one estimate indicated 30,000 episodes were available for download by listeners (Chung, 2008). Some recent blog conversations have suggested there may now be more than 300 million individual episodes, although these claims cannot be empirically verified. It is equally difficult to authoritatively gauge how many listeners are downloading ad using podcasts. Estimates have varied from 5 million (Huntsberger & Stavitsky, 2007), to 12 million (Bannan, 2006), to 8% of the European adult population –which would be approximately 17 million people (Klaassen, 2006). One estimate suggested 45 million people would be listening to podcasts by 2011 (Potter, 2006).

A podcast facilitates active information seeking (Inglis, Ling & Jootsen, 2001) because it puts the listener in control of the technology (Johnson & Grayden, 2006) Podcast creator can heighten the personalization of the medium through a "microtarget" approach (Snodgrass, 2007), meaning audio creators can make connections with listeners interested in highly specific ideas, entertainment, or products. Allowing the listeners to be in control of the communication while focusing on the specific needs and wants of that person creates the kind of marketing relationship organizations want today – one in which audiences and publics are aware, active, and assertive as consumers to get the information they need about relevant issues of life (Lesig, 2001).

Although college students as a whole are easily identified as technologically proficient media users, there is scant scholarly research addressing students' podcast use. The most noteworthy recent study involving students´ perception in general is (Chung 2008). Chung surveyed students to "explore the relationships between podcast use motivations and user behaviors and attitudes" Although valuable in terms of beginning a discussion of what prompts students to use and perceive value from podcasts, the study does not address issues of program preference or content sharing among listeners.

Likewise, a study by Parson, Reddy, Wood and Senior (2009) involved assessment of student opinions about podcasts –but only in the context of educational podcasts and the extent to which they added students' learning in class. There are a variety of other articles that also address the use of podcasts in the context of education (Copley, 2007; Evans, 2007; Huntsberger & Stavitsky, 2007; Richardson, 2006) or focus on student learning as it relates to podcast use (Jay, 2009; Lee, McLoughlin, & Chan, 2008; White, 2009).

Other published work deals with related technical or marketplace issues faced by podcast creators. Articles focus on means for producing effective podcasts (Zucker, 2008), the integration of podcasts into a marketing plan (Bannan, 2006; Klassen, 2006), or the means used by podcast creators to adapt to technological change (Green, 2007). These works and other similar articles are valuable in helping us understand how 'successful' podcasts can be created and used. However, the articles cannot predict what podcasts listeners will accept and their reasons for doing so.

The present study attempts to advance these earlier research efforts by profiling the extent to which a particular audience of technology adept listeners – college students – reported using podcasts, the program genres involved, the incorporation of listening with other daily activities, and listener's opinions about accessibility, entertainment value and other related variables. In an effort to do this, the study has been constructed within the perspective of uses and gratifications theory.

Uses and gratifications is an audience-centered approach. It views audiences as made up of goal-directed individuals who are active consumers of media (Katz, Gurevitch, & Hass, 1973). As these active individual consumers select media and content to attend to, consumers are motivated by their beliefs about how well the content will satisfy their social and psychological needs (Rayburn, Palmgreen, & Ackern 1984). The users and gratifications perspective views media use as an empowerment of individuals to make media choices for desired personal outcomes (Littlejohn, 1999).

Some media provide more opportunity for gratification than others (Greenberg, 1974). Users define gratification for themselves (Stafford, Stafford, Royne, & Schkade, 2004). This being the case, at least one researcher has found that a uses and gratifications theoretical model can be utilized to predict consumers' media choices (Dixon 1996).

The uses and gratifications model has been employed to help explain audience use of traditional print and broadcast media (Bryant & Zilman, 1984; Fisher, 1978; Massey, 1995) The perspective has also been used to help explain how audiences use online media content for entertainment, information gathering, and social interaction (Cowles, 1989; Papacharissi & Rubn, 2000). In recent years, uses and gratifications has been employed to help generate understandings about how users in a virtual media community have their "functional, emotive, and contextual needs" met by media content (Sangwan, 2005)

As with any audience group, college students do not readily accept all podcasts (Green, 2007). That being the case, it seemed appropriate to conduct research on the actions and opinions of college student podcast users within a theoretical perspective in which students could be seen as empowered individuals, active within a social group, making decisions about content use based on personal likes and dislikes, and then individually making choices about incorporating media use within daily activities.

3.   Methodology

This study was developed in collaboration with a group of 47 undergraduate students studying in ICSA at Jaurez City University a large North Mexican University. College students are active, aware, and technologically proficient in the use of digital audio programming (Parson et al., 2009); the author felt that student's involvement in the project would keep the research focused on issue most relevant to the survey population.

There are risks involved when utilizing students in research, because undergraduates often do not have acceptable conceptual and data collection skills (Berzonsky & Richardson, 2008). Even with the risks, the literature has also documented that faculty and undergraduates can successfully collaborate on research intended for publication (Lever, 2005). There has even a report of successful collaboration on grant-funded research between faculty and undergraduates with learning disabilities (Muller, 2006).

Faculty benefit from research collaboration by more efficiently reaching institutional research goals. Students benefit by gaining analytical skills that are valuable in the classroom and the workplace. One researcher recommended expanding faculty-undergraduate collaboration in the social sciences. (Ishiyama, 2002) wrote "collaborative research with faculty will undoubtedly assist in the effort to cultivate in students at liberal arts institutions the ability to engage in active learning".

Thorough, comprehensive faculty oversight is needed in every step of the process, and that was the case in this project. The author supervised all aspects of the students'work following the general guidelines used in an earlier peer-reviewed research project involving undergraduates (Swanson, 2004) so that professionalism and consistency would be maintained and reliable data generated.

The project began with an investigation of relevant theories, followed by a thorough review of literature. A total of 39 survey questions were developed and grouped into subject areas addressing participant demographics, access and use of digital media technology, podcast genre preferences, use of educationally oriented podcasts, and integration of podcasts into daily activities. Closed-ended and open-ended questions were developed. Ideally, an existing survey would have been used a guide in the development of the questionnaire, however, there is little literature dealing with podcast use (Brown, 2006) and no existing studies could be found that offered acceptable models for the questionnaire.

Questions were incorporated in an online survey that was approved by the university institutional human subjects committee. The questionnaire was made available on Survey Monkey between April 22 and June 22, 2010.

Respondents were targeted using a snowball sampling method. Snowball sampling is an approach for locating "information-rich key informants" (Effective engagement/snowball sampling 2007) such as those who would be likely to own a digital audio player and perhaps listen to podcast programming. Undergraduates assisting in the project disseminated a hyperlink to the online survey. Each e-mailed a link to the survey to his or her friends, or posted the link through social media that they use, such as Twitter or Facebook. Thirty two of the 47 students reported frequent use of Facebook; Fifteen of the students reported I excess of 700 Facebook friends.

During the 90-day survey period, 286 surveys were returned. Of these, 63 were disregarded due to incomplete survey response. A total of 223 complete responses comprised the sample for study. Information provided by these respondents allowed for the development of answers to four research questions focusing on college student listeners' use of podcasts and opinions about podcast content and value.

4.   Research questions

RQ1: To what extent do college students report downloading, listening to, and sharing digital audio podcasts?

RQ2: What podcast program genres are most popular among college student listeners?

RQ3: How do college students incorporate podcast listening into their daily activities?

RQ4: What are college student listeners' opinions about accessibility, technical quality, usefulness and entertainment value of podcast programming?

5.   Results

Descriptive statistics indicate that the average age of respondents was 21.9. Among all respondents, 96% identified as undergraduate students and 4% identified as graduate students. Slightly more than half of all respondents (53%) reported being enrolled in college for two years or less, and 41% reported three to five years of enrollment. Five percent of respondents reported more than five years of college enrollment.

Respondents were asked to identify their home state or country. Among those reporting, a total of 12 Mexican states were identified (99% of total response) with two respondents indicating Brazil as their country of residence.

Respondents were asked to report their major field of study. Among those reporting the results were broadcasting/journalism/communication (49.3%), humanities/language/literature/social sciences (18.8%), architecture, engineering or applied sciences (9.9%), science or math (7.2%), computer science/information technology (5.4%), music/theatre/fine arts (4%), agriculture (3.1%), athletics/physical education/sports medicine (2.2%).

RQ1: To what extent do college students report downloading, listening to, and sharing podcasts?

All respondents reported owning at least one portable audio player (166, or 74%) and about one in four respondents (57 or 26%) reported owning more than one device. Apple's iTunes was respondents' preferred source for podcasts, with 82% (182) citing iTunes as their top download source, followed by podcast host web pages (48, or 21%), Google (30, or 13%) and other search engines (8, or 3%).

When asked how often they had downloaded a podcast episode during the current year, the majority of respondents reported one or fewer downloads per month (128, or 58%). Forty-two respondents (19%) reported downloading two or three times a month. Forty-nine respondents (22%) reported downloading at least one episode per week.

The vast majority of respondents reported downloading only one or two different podcast programs (173, or 78%9, and 47 respondents (21%) reported downloading between three and 10 different programs. A much smaller number of respondents (3, or 1%) claimed regular download of more than 10 programs.

Respondents' estimate of time spent listening to podcasts seems consistent with reports of program downloading. More than half of all respondents reported listening to podcasts for an hour or less per month (114, or 51%). About one-third of respondents reported listening for five or more hours per month (36, or 16%).

Respondents were asked if any of their college instructors had offered podcasts as part of course content, and, if so, whether the podcasts had been used by students. Most respondents (130, or 58%) said none of their instructors had ever used podcasts. Although 93 respondents (41%) had taken courses that included podcast content, only 36 (16%) reported that they regularly listened to those podcasts.

For many respondents, podcast listening is a solitary activity. Although slightly more than half of respondents reported "occasionally" sharing programs or program information (115 or 51%), 80 respondents (35%), "never" share. Twenty-eight respondents (12%) reported they "regularly" or "always" share with others.

RQ2: What podcast program genres are most popular among college student listeners?

Respondents were asked what types of podcasts they most commonly listen to fifteen category options (consistent with categories used in the iTunes directory) were given Respondents´ preferences are shown in Table 1.

Table 1. Respondents' reported podcast listening preferences, by genre (multiple individual selections allowed). 223 respondents offered 608 preferences (average of 2.73 per respondent).

News and politics	84	38%
Comedy	79	35%
TV, film, popular culture	71	32%
Music	68	30%
Education	57	26%
Religion and spirituality	44	20%
Technology	35	16%
Arts, literature	33	15%
Sports and recreation	31	14%
Business	24	11%
Science and medicine	24	11%
Government	21	9%
Health	17	8%
Games and hobbies	13	6%
Kids and family	7	3%

Results show respondents download more program episodes than they listen to. About one in four listeners reported listening "regularly" or "without fail" to episodes they downloaded (52 or 23%). Remaining respondents listeners said they "always" listen to the entire contents of podcast episodes downloaded (98, or 44%), ad an even smaller number of respondents listening to episodes more than once (100, or 45%).

RQ3: How do college students incorporate podcast listening into their daily activities?

Results confirm that college students are technologically capable consumers. In other words, they frequently use a computer to access the Internet and download media content. The vast majority of respondents (206, or 92%) report spending at least 3 hours per week with a computer or other digital information device. More than half of all respondents (117, or 52%) agreed with the statement "The Internet is a constant, essential presence in my life.

Respondents were asked what they do when listening to podcasts. Results are shown in Table 2.

Table 2. Activities that respondents report participating in while listening to podcasts (multiple individual selections allowed). 223 respondents offered 463 preferences (average of 2.08 per respondent).

I listen while reading or studying.	109	49%
I listen while cleaning or doing housework.	87	39%
I listen when driving my car.	67	30%
I listen when walking or bicycling.	60	27%
I listen while exercising.	53	24%
Other.	45	20%
Elaborative comment required, see below		
I listen when riding public transportation.	42	19%
I listen while completing other tasks on the computer.	14	6%
I listen while I am at my job.	3	1%

Note. Elaborative comments included:"I listen while doing nothing else".."I do not participate in any other activity".."Just sitting on the toilet".."When I am hanging out".."When I am relaxing at home".."While eating".."In bed".."When I have free time".."While taking study breaks".."I listen while cooking".."Just sit and listen".."While playing video games".."While laying in bed before going to sleep".

RQ4: What are college student listeners' opinions about accessibility, technical quality, usefulness, and entertainment value of podcast programming?

More than two-thirds of residents reported that they most often listen to podcasts through their computer than through a portable device (152, or 68%). When asked the type of computer used most often, 120 respondents (54%) reported using a PC and 103 respondents (46%)reported using a Mac.

Respondents' opinions about the appropriate length of a podcast episode varied widely. A slight majority of respondents (75, or 34%) indicated 15-30 minutes was the ideal length, and a lesser number (58, or 26%) indicated a preference for 1015 minutes episodes. Forty-seven respondents (21%) preferred podcast episodes of less than 10 minutes. Forty-three respondents (19%) preferred episodes of 30 minutes or more.

A series of questions was presented to gather respondents' opinions about accessibility, technical quality, usefulness, and entertainment value of podcast programming. Results are shown in Table 3.

Table 3. Respondents' reported opinions about accessibility, technical quality, usefulness, and entertainment value of podcast programming $n = 223$.

		Agree or strongly agree	Unsure	Disagree or strongly disagree
Accessibility	Finding and downloading podcasts is a big hassle.	19% (42)	18% (41)	63% (140)
Technical quality issues	The technical quality of podcast programming is better than radio.	67% (148)	27% (61)	6% (14)
Entertainment value	I look forward to listening to new podcasts and podcast episodes.	51% (115)	33% (73)	16% (35)
	Podcasts offer me information and/or entertainment that I could not get anywhere else.	46% (102)	25% (56)	29% (65)
	Most podcasts are just trying to sell me something.	8% (21)	22% (50)	68% (152)
Usefullness	It's fun to listen to podcasts and then share what I heard with my friends.	49% (110)	35% (77)	16% (36)
	Most episodes I listen to are fairly repetitive (they sound like the last episode of the same podcast).	25% (56)	41% (91)	34% (76)
	I seldom learn anything new when listening to a podcast.	12% (26)	18% (40)	70% (157)

6.   Discussion

The study findings present some interesting and heretofore unreported insights into college students' opinions about the usefulness and value of podcast programming. The findings also present an interesting profile of respondents´ reported downloading and use of podcasts. At the same time, some of the data seem inconsistent and perhaps contradictory.

This study found that while college students frequently affirmed that the Internet is "a constant, essential presence" in their lives; respondents generally were not heavy users of podcast programming found on the Internet. More than half of the respondents of this survey (51%) reported spending an hour or less per month listening to podcasts, and the majority of respondents reported downloading a podcast no more than once a month. As the reported frequency of podcast downloading increased, the percentage of students engaged in downloading dropped off markedly. Only a handful of respondents reported downloading more than once a week

Although fewer than half of respondents reported spending in excess of an hour a month listening to podcasts, a large number of respondents (78%) reported that they regularly download no more than two podcast programs. Few students reported they were profligate downloaders; only three individual respondents reported regularly downloading more than 10 programs.

The fact that most students reported so little activity (in terms of program download and listening time) calls into question respondents' claimed preference for podcast programs that are 15 minutes or more in length. (A program length or 15 minutes or more was identified as ideal by 53% of respondents). The reported lack of the activity also calls into question respondents' claims that they sample widely among the 15 genres of podcast programming presented in the survey (Table 1). Respondents made a total of 608 selections of "listened to" genres. This means that on average, each respondent identified an average of 2.7 favored genres. The numbers do not seen to be consistent. Would listeners who rarely spend more than an hour a month downloading programming, do not typically download more than a few different programs, and prefer programs in excess of 15 minutes in length, reasonably identify so many favorites across a wide spectrum of genres? It seems unlikely.

The data gathered by the present study also suggests that some respondents may not adequately have understood what was being asked of them. This is a common problem in survey research (Babbie, 1990) The survey instructions informed participants three times that the survey was aimed at obtaining the opinions of college students who have "ever listened to an audio podcast (digital audio downloaded via the Internet")". Despite the presence of these instructions and a request to answer all questions, 63 surveys were submitted incomplete and were subsequently disregarded during the initial screening process. Several of the surveys that were disregarded included comments suggesting that the respondents did not read or understand the initial instructions. One comment was: "You could also have options for people who don't listen to podcasts". Another was: "I have never downloaded

One might surmise from the findings that when it comes to podcast program listening, many college students have greater ambition than they do follow through. About half of all respondents reported that they "occasionally" listen to podcasts they download, and almost one out of every four respondents reported "I never listen" to the programs downloaded. Likewise, fewer than half of all respondents answered affirmatively when asked if they "listen to the entire contents of the podcast episodes that you download".

Respondents indicated that they engage in a variety of activities when listening to podcasts (Table 2). At the same time, most respondents (152, or 69%) reported listening to podcasts "mostly through my computer". As a result, it might be difficult to conclude that podcasts are widely used 'on the go' −at least by many of the student who respondents to this survey.

Respondents reported that the technical quality of podcast programming is better than radio. Respondents agreed that podcasts are easy to access and provide information and entertainment that listeners could not find anywhere else. Respondents affirmed that they often learn something new when listening to aa podcast program, and that they do not perceive podcasts as "just trying to sell me something".

Students reported that they look forward to listening to new podcasts and sharing what they have heard with their friends, although there was no clear consensus on the question of whether podcasts "of my favorite program (are) made available for download as often as I would like".

As previously noted, respondents provided a wealth of information about the perceptions of technical quality, usefulness and entertainment value of podcast programming. All of this is information not gathered by previous researchers.

7.   Limitations

The small sample size and the choice of sampling methodology make it problematic to use the findings of this research to reach broad generalizations about the opinion and behaviors of college students as a whole. Ideally future research should be conducted using a much larger sample of college students and perhaps stratified across class standings (freshman, sophomore, junior, senior, and graduate students) and demographics groups (age, sex, racial/ethnic background, in others).

Snowball sampling was chosen as a quick, easy method of gathering data. The tradeoff for expediency is the lack of a random sample. By design, the chosen methodology resulted in the non-random acquisition of data from participants who were social networking ´friends´ of the communication students assisting the author with the project. Consequently and perhaps not surprisingly, almost half the respondents (110 or 49%) were majoring in communication fields. Future research should attempt to acquire a large and random sample of college students from different academic program areas.

Although it was certainly advantageous to have responses from students who were familiar with the technology they were being asked about, it is unclear whether the results of the study would have differed if the subject population were more evenly distributed across the academic disciplines. At the same tie, it could also be argued that since respondents targeted for participation in the survey were already involved in social media (through links to the students who began the 'snowball' sample), respondents could have had more exposure to podcasts and a greater propensity to have strong opinions about this media content than students not involved in social media.

Two-thirds of students who reported they were exposed to podcasts associated with college courses said they never listened to those programs. Future research could help determine if lack of attentiveness to course podcasts is widespread, and the reasons behind students' lack of attentiveness.

Under the uses and gratifications perspective, it is "an established notion" that personality differences among audience members may relate to the media programming choices made by those audience members (Bagdasarov et al., 2010). In the context of the present study then, it could be speculated that certain college students may be predisposed to 'tune in' to certain types of podcast programming. The small sample size does not allow confirmation of this supposition, and the apparent flighty nature of student podcast sampling noted in table 1 might suggest otherwise.

Different forms of computers-mediated communication bring about different practices by users. (Quan-Haase and Young, 2010) suggested, for example, "Facebook is about having fun and knowing about the social activities occurring in one's social network, whereas instant messaging is geared more toward relationship maintenance and development". In this context, then, podcasts would not allow immediate, electronic interaction among users. Further investigation within uses and gratifications perspective would address the different social practices of users and how those practices interact with specific types of social media.

Although this author felt that the uses and gratifications perspective was the ideal theoretical framework under which to conduct a study of this type, others would disagree. Ruggiero contended that applying uses and gratifications makes "makes it difficult to explain or predict beyond the people studied or to considered societal implications of media use" (Ruggiero, 2000). Indeed, given what could be interpreted as inconsistencies or inaccuracies in response among those who participated in this survey – this is a valid concern in research of this type.

The author and collaborating students worked diligently to develop a questionnaire that was clear ad concise. Detailed but simply worded instructions were offered. The instrument was subject to extensive review prior to approval by the university's human subjects committee. At the same time, it would have been idea to base the questionnaire or an instrument previously used in similar research. No such instrument could be found.

Twenty-two percent of respondents who attempted the survey did not answer all the questions. Despite instructions clearly identifying the survey as focusing on podcast users only, many incomplete responses came from respondents who started they did not use podcasts. Although the author believes all reasonable precautions were taken to construct a valid and reliable instrument, further review of the instrument and the instructions would seem worthwhile.

Even with incomplete responses were weeded out, the question remains whether the responses of those participants who did indicate familiarity and involvement with podcasts are accurate in terms of respondents' opinions and behaviors. Some respondents' answers may be more reflective of 'wishful thinking' than 'actual behavior'. There is no way to tell.

8.   Conclusion

This study involved a small sample of college students, to gather information about their use of, and opinions about, podcast programming. The study raises issues that have been unaddressed thus far I the literature, including: What kinds of programming do college students listen to, how do students integrate podcast listening and sharing into their lives, and, what do collegiate listeners think about the technical quality, usefulness and entertainment value of podcast programming? These questions have not been asked before, and the answers to them (even on a small scale) are worth our attention.

Although most respondents agreed with the statement "The Internet is a constant, essential presence in my life", most respondents completed one or fewer podcast downloads per month. More than half reported spending an hour or less per month listening to podcasts. Twenty-two percent of respondents said they "never" listen to podcasts they download. At the same time, the results to the survey items about quality and value of podcast programming seem to suggest students enjoy the programs they do download and listen to.

College students are, as a whole, quite comfortable with new technology and have a huge impact on the consumer marketplace. The author hopes much more the scholarly community will do investigate work so that we can understand the many ways college students impact and are impacted by podcast audio programming.

References

Babbie, E. R. (1990), Survey research methods (2nd ed.), Belmont, CA: Wadsmorth.

Bagdasarov, Z,, Greene, K., Banerjee, S. C., Krcmar, M., Yanovitzky, I., & Rugintye, D. (2000), I am what I watch: Voyeurism, sensation seeking and television viewing patterns Journal of broadcasting & Electronic Media, 54(2), 299-315, doi:10.1080/08838151003734995.

Bannan, 2006 K.J. Bannan Using podcasts to build brands BtoB Magazine Online (2006)   Retrieved   from www.btobonline.com/apps/pbcs.dll/article?AID=20060403/FREE/604030707&templa te=printart

Berzonsky, W. and Richardson, K.   2008, Referencing science: Teaching undergraduates to identify, validate, and utilize peer-reviewed online literature Journal of Natural Resources and Life Sciences Education, 37 (2008), pp. 8–13 Retrieved from Education Full Text database

Brown, D. (2006). Generation iPod: An exploratory study of podcasting's 'innovators.' Paper presented at the annual meeting of the Association for Education in Journalism and Mass Communication, San Francisco, August, 2006.

Bryant, J. & Zillman, D. (1984) Using television to alleviate boredom and stress. Journal of Broadcasting, 28 (1984), pp. 1–20

Bullis, K. (2005). Podcasting takes off. Technology Review (2005, October), p. 30.

Chung, M. (2008). Podcast use motivations and patterns among college students. Unpublished master's thesis. Manhattan, Kansas: Kansas State University.

College students setting records in spending (2008, April). Civic engagement, digital connectivity,   2008.   Alloy   Media + Marketing.   Retrieved fromhttp://www.marketingcharts.com/television/college-students-setting-records-in-spending-civic-engagement-digital-connectivity-5533/alloy-college-company-positive-impact-criteria-2008jpg

Copley, J. (2007) Audio and video podcasts of lectures for campus-based students: Production and evaluation of student use Innovations in Education & Teaching International, 44 (4) (2007), pp. 387–399

Couch, C.J. (1996). Information technologies and social orders Aldine de Gruyter, Hawthorne, NY (1996)

Cowles, D. (1989). Consumer perceptions of interactive media. Journal of Broadcasting & Electronic Media, 33 (1989), pp. 83–89

Demaria, M. (2004). Combat user ignorance. Network Computing, 15 (26) (2004), p. 65

Dixon, J. (1996). **Uses and gratifications theory to predict use of computer mediated communications.** International Journal of Educational Telecommunications, 2 (1) (1996), pp. 3–27.

Effective engagement/snowball sampling (2007, July 11). State of Victoria Department of Sustainability and Environment. Melbourne, Victoria, Australia. Retrieved from http://www.dse.vic.gov.au/dse/wcmn203.nsf/linkview/d340630944bb2d51ca25 708900062e9838c091705ea81a2fca257091000f8579

Evans, C. (2007). **The effectiveness of m-learning in the form of podcast revision lectures in higher education.** Science Direct (2007, September 25), pp. 491–498

Fisher, B.A. (1978). **Perspectives on human communication.** MacMillan, New York (1978)

Green, H. (2007). **Don't quit your day job, podcasters.** Business Week (2007, April 9), pp. 72–74

Greenberg, B.S. (1974). **Gratifications of television viewing and their correlates for British children.** J.G. Blumler, E. Katz (Eds.), The uses of mass communication: Current perspectives on gratifications research, Beverly Hills, CA, Sage (1974), pp. 71–92

Huntsberger, M. & Stavitsky, A. (2007). **The new 'podagogy': Incorporating podcasting into journalism education.** Journalism & Mass Communication Educator, 61 (4) (2007), pp. 397–410

Inglis, A.; Ling, P. & Jootsen, V. (2001). **Delivering digitally: Managing the transition to the knowledge media.** (2nd ed.)Kogan Page, London (2001)

Ishiyama, J. (2002). **Does early participation in undergraduate research benefit social science and humanities students?.** College Student Journal, 36 (3) (2002), pp. 380–386 Retrieved from Education Full Text database.

Jay, C. (2009). **Students who use podcasts fare better on tests.** The Daily Skiff (2009, February 27) Retrieved from http://media.www.tcudailyskiff.com/media/storage/paper792/news/2009/02/27 /News/Study.Students.Who.Use.Podcasts.Fare.Better.On.Tests-3651534.shtml

Johnson, L. & Grayden, S. (2006). **Podcasts – An emerging form of digital publishing.** Journal of Computerized Dentistry, 9 (3) (2006), pp. 205–218.

Katz, E., Gurevitch, M. & Hass, H. (1973). **On the use of mass media for important things**

American Sociological Review, 38 (2) (1973), pp. 164–181.

Klaassen, A. (2006). **Reality check: Blogs and ipods are potent tools, but the reach most marketers crave still comes from good 'ol TV, print and Internet ads.** Advertising Age (2006, August 31), p. 1

Lee, M. McLoughlin, C. & Chan, A. (2006). **Talk the talk: Learner-generated podcasts as catalysts for knowledge creation.** British Journal of Educational Technology, 39 (3) (2008), pp. 501–521

Lesig, L. (2001). **The future of ideas: The fate of the commons in a connected world.** Random House, New York (2001)

Lever, K. (2005). **Introducing students to research: The road to success.** Journal of Nursing Education, 44 (10) (2005), pp. 470–472 Retrieved from Education Full Text database.

Littlejohn, S. W. (1999). **Theories of human communication.** (6th ed.)Wadsworth, Belmont, CA (1999).

Marsal, K. (2006). **iPod: How big can it get?.** AppleInsider (2006, May 24) Retrieved from http://www.appleinsider.com/article.php?id=1770.

Massey, K.B. (1995). **Analyzing the uses and gratifications concept of audience activity with a qualitative approach: Media encounters during the 1989 Loma Prieta earthquake disaster**

Journal of Broadcasting & Electronic Media, 39 (1995), pp. 328–342

Muller, L. (2006). **Research collaboration with learning-disabled students.** Journal of College Science Teaching, 36 (3) (2006), pp. 26–29 Retrieved from Education Full Text database

Notess, G. (2005). **Casting the net: Podcasting and screencasting.** Online, 29 (2005, November/December), pp. 43–45.

100 million iPods sold (2007, April 9). News release. Cupertino, CA: Apple Corporation. Retrieved from http://www.apple.com/pr/library/2007/04/09ipod.html

Papacharissi, Z. & Rubin, A.M. (2000). **Predictors of Internet use.** Journal of Broadcasting and Electronic Media, 44 (2) (2000), pp. 175–196

Parson, V., Reddy, P. & Wood, J. (2009). **Educating an iPod generation: Undergraduate attitudes, experiences and understanding of podcast and podcast use format.** Learning, Media, & Technology (2009, September), pp. 215–218

Potter, D. (2006). **iPod, you pod, we all pod: Eager to lure news consumers, media outlets are experimenting with news-on-demand podcasts.** American Journalism Review (2006, February/March), p. 64.

Quan-Haase, A. & Young, A. (2010). **Uses and gratifications of social media: A comparison of Facebook and instant messaging.** Bulletin of Science, Technology & Society, 30 (5) (2010, August 30), pp. 350–361.

Rayburn, J.D.; Palmgreen, P. & Acker, T. (2009). **Media gratifications and choosing a morning news program.** Journalism Quarterly, 61 (1984), pp. 141–156

Richardson, W. (2006). **Blogs, wikis, podcasts, and other powerful tools for classrooms**

Corwin Press, Thousand Oaks, CA (2006)

Ruggiero, T.E. (2000). **Uses and gratifications theory in the 21st century.** Mass Communication and Society, 31 (1) (2000), pp. 3–37.

Sangwan, S. (2005). *Virtual community success: A uses and gratifications perspective.* Paper presented at the 38th Annual Hawaii International Conference on System Sciences, January. Retrieved from http://www2.computer.org/portal/web/csdl/doi/10.1109/HICSS.2005.673

Snodgrass, M. (2007). Researching podcasts: Here's how to do it. *PR News Digital PR Report*. Retrieved from http://prnewsonline.com/digitalpr/howto/researching_podcasts.html

Stafford, T.F.; Stafford, M.R. & Schkade, L.L. (2004). **Determining uses and gratifications for the Internet.** Sciences Atlanta, 35 (2) (2004), pp. 259–288

Survey: Apple iPod bigger than beer among college students (2006, June 8). Retrieved from www.macdailynews.com/index.php/weblog/comments/9811/.

Swanson, D.J. (2004). **Homeland attack and homepage response: A preliminary review of affected business entities' online corroboration/explanation of September 11 economic impact**

The Social Science Journal, 41 (2) (2004), pp. 301–307

The Digital Disconnect (2002, August 14). Pew Internet & American Life Project. Retrieved from http://www.pewInternet.org/reports/toc.asp?Report=67

Tips for podcast fans (2009). Apple Inc. Retrieved from http://www.apple.com/itunes/podcasts/

U.S. Department of Education, National Center for Education Statistics. (2009). *Digest of Education Statistics*, 2008 (NCES 2009-020).

White, B. (2009). **Analysis of students' downloading of online audio lecture recordings in a large biology lecture course.** Journal of College Science Teaching, 38 (3) (2009), pp. 23–27

Zucker, M. (2008). **It's so easy to produce a podcast, there's really no excuse not to**

Advertising Age (2008, May 5), p. 18

Translating Natural Language Queries in Spanish to SQL involving Group By

[1]Andrés Bautista, [1]José Martínez, [2]Alberto Ochoa-Zezzatti

[1]Instituto Tecnológico de Ciudad Madero
{andres,bautista@live.com.mx, jose.mtz@itcm.edu.mx}

[2]Juárez City University
{alberto.ochoa@uacj.mx}

Abstract. This paper describes the analysis carried out in the translation of Natural Language Queries in Spanish to SQL involving the clause of grouping GROUP BY in Natural Language Interfaces to Databases (NLIBDs), the important role and the different ways to find them in the Natural Language.

Keywords. Aggregate Functions, Natural Language Interfaces, Natural Language Processing.

1   INTRODUCTION

Currently the majority of the information stored in databases (BD), is subsequently consulted for decisions make. To facilitate consultation of information in the databases have developed several tools that allowed the easy job of users (e.g. consultations assistants, graphical interface with menus, etc.), Many of the tools developed can generate queries information to meet the requirements of users, however can't perform any type of query because of its limitations, for this reason we developed the Natural Language Interfaces to Databases (NLIBD) through which we can get the information from a BD with a natural language query [4].

Some of these queries containing statistical expressions equivalent to processing the data stored in one or more tables, through aggregation functions and the GROUP BY clause. With SQL data can be grouped and added so that users can interact with them on a higher level of granularity, as stored data in databases.

For that an ILNBD provide the information requested in a query, this must know oral or written expression of the people, which communicate in Spanish Natural Language, the processing queries starts with a lexical analysis and finish at the time of generate the SQL query.

The ILNBD have developed since the 60s and unfortunately not generated a 100% of correct answers to queries provided by users, this is mainly because most of ILNBD do not have the ability for processing queries involving aggregate functions or grouping [2].

This article is a description and analysis of queries involving aggregation and grouping functions, showing examples and reviewing the process necessary for the correct translation into SQL form.

2   Natural Language Interface

The natural language processing (NLP) is a set of computational techniques to analyze and represent texts naturally in one or more levels of linguistic analysis, in order to carry out the processing of language as a human for a range of tasks and applications [1].

Natural language interfaces are mechanisms of communication between persons and a machine through natural language. Typically, this communication is bidirectional, (i.e. question-answer type). The general architecture of an ILN is shown in Figure 1.

User

57

Figure 1. General Architecture of ILN

3   Databases

A database is a collection of related data. With the word data we refer to facts (data) known recordable and have an implicit meaning [2].

A databases have the followings implicit properties.

•   A database represents some aspect of the real world, sometimes called mini-world or universe of discourse.

•   A database is a collection of logically coherent data, with some inherent meaning.

•   The database is designed, built and tested with data for a specific purpose.

The computerized databases can create and maintain with a group of application programs written specifically for those tasks, or by a management system database.

A database management system (DBMS) is a software that allows at users to create and maintain a database. Therefore, the DBMS is a software system which facilitates the general purpose process to define, built and manipulate databases for various applications.

4   Aggregate Functions and GROUP BY clause

Aggregate functions are functions that take a collection of values as input and produce a single output value. SQL provides five primitive aggregation functions:

1. **COUNT:** returns the total number of rows selected.
2. **SUM:** Adds the values of a column.
3. **MIN:** returns the minimum value of a column.
4. **MAX:** returns the maximum value of a column.
5. **AVG:** Calculate the average value of a column.

In addition to expanding the use of aggregation functions is necessary to use of **GROUP BY** clause, which used to group rows by specific columns. [3].

5   Natural Language Interface to Databases

The figure 2 shows the flow of NLIDB, in which the result is usually presented in two ways, as in SQL statement or as an answer in natural language. In this article the results are returned as SQL language instruction.

Some major NLIDB founded in the literature that have been developed are described in Table 1, further noting the use of aggregation functions [4].

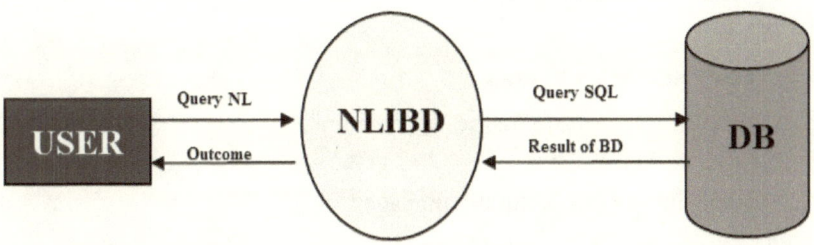

Figure 2. NLIDB Flow

Interface	Aggregate Functions
TAMIC (1996)	x
IDICULA (1999)	x
PRECISE (2003)	x
InBase (2003)	x
NLPQC (2005)	x
Translator CENIDET (2005)	x
WYSIWYM (2006)	x
Translator OWDA Dravidian Language (2007)	✔
C-PHRASE (2008)	x
Translator Rojas (2009)	x
STK (2010)	x
Translator Esquivel (2010)	✔
Current job ITCM (2012)	✔

Table 1. Main NLIBDs developed.

6   Analysis of Translation of aggregate functions and GROUP BY clause.

As we have seen in section 4, the aggregation functions allow us to perform operations on the information to be able get a better result in our queries to databases.

To better understand the use of aggregate functions, then show the syntax they use.

- **MAX and MIN Syntax:**

SELECT MAX/MIN ("name_of_column")
FROM "name_of_table"

Example in Spanish Natural Language:

"Dame el precio mayor de los productos".

(Give me the higher price of the products)

The SQL sentence generated is:

SQL: SELECT MAX(precio)

FROM PRODUCTOS

- **SUM Function and GROUP BY clause:**

SELECT "name1_column", SUM ("name2_column")
FROM "name_table"
GROUP BY "name1-column"

Example in Spanish Natural Language:

"¿Cuantos trabajadores hay en cada departamento?"

(How many employees are there in each department?)

The SQL sentence corresponding is:

SQL: SELECT department, count (employee)

   FROM departments, employees

   WHERE departments.id = employees.idDepartment

   GROUP BY department

As shown in the above examples use of aggregation functions allows the user to get more specific information.

As shown in the above examples use of aggregation functions allows the user to get more specific information.

If aggregate functions are as necessary and extensively used in the real world, what is the cause which prevents the implementation of so useful recovery options in ILNBDs information? To get the answer to this question is needed extensive analysis on translation techniques for each NLINDBs developed. But as we talked Natural Language Spanish, we can see some patterns that sentences or phrases used in the queries that are made on ILNBDs indicating use of aggregate functions and GROUP BY clause.

To understand the above we have focused on the analysis of some queries of the corpus of the Linguistic Database Cultures of the World: A Statistical Reference, an adaptation of Philip M. Parker. The BD mentioned has only two tables (social_demography, geography), where it is concentrated the information of the linguistic groups of the world, its geography, demographics, etc.

Examples of querys of the corpus:

1.   Sociedades que viven en clima templado (Societies that live in temperate climate).
2.   Nivel de Deforestación (Doforestation level).
3.   Mayor número de fronteras(Greater number of forontiers).
4.   Mayor   ocurrencia   de   Terremotos   (Major   occurrence   of earthquakes).
5.   Clasificación de Sociedades por Huso Horario (Society clasification by TimeZone).

Query 1 should be resolved properly in any of the ILNBDs and generates a SQL statement similar to the following:

SELECT society
FROM geography
WHERE climate='templado'

For query 4 the response of ILNBDs most current, if give any response, would be omit the word 'classification' and show the occurrence of earthquakes in all societies, the SQL translation would be the following.

SELECT earthquakes
FROM geography

Some ILNBDs are adaptive and may add new patterns of recognitions of sentences, but would imply add a new pattern for each type and structure of oration that can be formed in our extensive Spanish Natural Language, increasing the use of resources needed for processing.

What happens if the query 3 is introduced in other ways? Examples:

— Dame el número mayor de fronteras (Give me the largest number of frontiers).
— Muéstrame el mayor número de fronteras (Show me the greater number of frontiers).
— De las fronteras ¿cuál es el número mayor? (Of frontiers what is the largest number?).

When the query is analyzed in detail we note that the degree of difficulty to understand, through a language translation as those used in ILNBDs increases, which is why deserve aggregation functions will be analyzed from different points of view, before he could speak its implementation.

Returning to the example of the query 4, we note that the user is requesting only one fact, the higher occurrence of earthquakes that have registered, to solve this query and not shed excess information or erroneous through the MAX aggregation function can show data that the user requests, see the equivalent query in SQL.

SELECT MAX(earthquakes)
FROM geography

As we can see the grouping is very important, although it is clear that the examples are very simple due to the database that was used, but in companies where information is concentrated part of a large network of department stores and all information is stored in a single BD, means developing a complex analysis of the query that is being requested and include the necessary relationships, we see this with an example query in Spanish Natural Language:

"Dame el número de trabajadores del departamento de carnes de las sucursales de la ciudad de México que tengan menos de 2 años de antigüedad".

(Give me the number of workers in the meat department of the branches of Mexico City with less than 2 years of antiquity)

To solve the above query is first necessary to determine the relationship, in this case, conjoined entities to obtain the necessary information are empleados, departamentos, sucursales y ciudades (Employees, departments, branches and cities), then we have to consider whether need to use aggregate functions or Group By clause, for this case are necessary both, the SUM aggregate function to count the number of workers and the clause to group by department.

This article is the beginning of the development of a master project that is planned in the Technological Institute of Ciudad Madero (ITCM), which aims to solve the translations of Spanish natural language queries on relational database to extend on a translation domains ILNBD.

Some of the keywords that we will be analyzing when translating queries to identify the use of aggregate functions are shown in Table 2, in a column that is the word in Spanish and another NL aggregation function corresponding remembering that only show the main, however, the number will increase in our implementation because they consider all possible synonyms that exist in the Spanish LN and words or phrases that may arise.

Palabra/Frase	Función de Agregación/Cláusula
Cuantos	COUNT
Suma	SUM
Promedio	AVG
Media	AVG
Máximo	MAX
Mayor	MAX
Mínimo	MIN
Menor	MIN
Todos los(as)	COUNT
El Total	SUM
Agrupado	GROUP BY
Clasificado	GROUP BY

Table 2. Word analysis to use aggregate functions and GROUP BY clause.

7   CONCLUSION

As we have seen throughout this article, the development of natural language interfaces that translate queries involving aggregate functions and GROUP BY clause requiring a good discussion and good solution strategy to allow correct translation query to be processed by the interface.

Resolve issues important to natural language processing and applying them in NLI, enhances the domain of information that can be obtained any relational DBs.

BIBLIOGRAPHY

Liddy D, *Natural Language Processing for Information Retrieval & Knowledge Discovery, School of Information Studies*, 2001.

Androutsopoulos I., Ritchie G.D., Thanisch P. *Natural Language Interfaces to Databases - An Introduction. Natural Language Engineering*, 1995

Carme Martín Escofet, *El lenguaje SQL.* Rojas J.C. *Administrador de Diálogo para una Interfaz de Lenguaje Natural a Bases de Datos,* 2009.

Elmasri and Navathe, *Fundamentos de Sistemas de Bases de Datos,* Ed. Pearson, 5 Edition, 2007.

Intelligent System for Diagnosis Hypertiroidism using Case-based Reasoning

Alberto Ochoa[1,*], Rubén Jaramillo[2], Sandra Bustillos[3], Daniel Azpeitia[1], and Saúl González

[1] Juarez City University, México

[2] CFE-LAPEM.
alberto.ochoa@uacj.mx

Abstract

The process of medical diagnosis is always complex. It demands the valuation of multiple interacting factors in the case under examination.

The signs and symptoms of the patient are put under the experienced opinions of one or more doctors whom propose a corresponding treatment. With the advent of Artificial Intelligence (AI) techniques such as Case-based reasoning (CBR), education of medical subjects has become effective. CBR plays an important role in building intelligent system for disease prognosis and diagnosis. We discuss a CBR-based intelligent system built for diagnosis Hyperthyroidism. The results of the proposal system justify its usefulness for this sickness which suffers 18750 people in Mexico.

Keywords: Hyperthyroidism, Case-based reasoning, Medical Education.

1 Introduction

Last years of the previous century and the first decades of the next one, indicate a tendency in the evolution of the research n the field of the health. The superior medical education, in its constant improvement, requires the introduction of technical outposts to prepare an individual able to stay updated in its specialty during all his or her life. National System of Mexican Health has developed a plan of action for increasing the quality of people's life by continuously updating the health sector professionals through innovative medical education systems. Human resource and its development are always important for a national growth. In the modern era, computer systems are an important ally for this purpose. Healthcare forms an integral part of improving quality of human life Integration of modern intelligent systems for medical education plays a significant role.

2. Hyperthyroidism.

Hyperthyroidism, often referred to as an 'overactive thyroid', is a condition in which the thyroid gland produces and secretes excessive amounts of the free (not protein bound and circulating in the blood) thyroid hormones, triiodothyronine (T3) and/or thyroxine (T4). This is the opposite of hypothyroidism ('sluggish thyroid'), which is the reduced production and secretion of T3 and/or T4. Hyperthyroidism is a type of thyrotoxicosis, a hypermetabolic clinical syndrome which occurs when there are elevated serum levels of T3 and/or T4. Graves diseases is the most common form of hyperthyroidism.

While hyperthyroidism may cause thyrotoxicosis they are not synonymous medical conditions; some patients may develop thyrotoxicosis as a result of inflammation of the thyroid gland (thyroiditis), which may cause the release of excessive thyroid hormone already stored in the gland but does not cause accelerated hormone production. Thyrotoxicosis may also occur by the ingestion of excessive amounts of exogenous thyroid hormone in the form of thyroid hormone supplements such as the most widely used supplement levothyroxine, liothyronine, in weight-reducing dietary supplements that contain thyroid hormone, synthetic forms of T4 and T3 or thyroid extract (dessicated thyroid). Excessive exogenous intake may be purposeful as part of various treatment regimens such as to suppress tumor growth in thyroid cancer or inadvertently, as in when using dietary supplements or via percutaneous absorption, as a result of topical use of cosmetic creams containing iodine or thyroid hormones. In these cases it is termed *Thyrotoxicosis Factitia* (*facticius* :artificial or self-induced); it is also known by other terms such as exogenous thyrotoxicosis, alimentarythyrotoxicosis or occut factitial thyrotoxicosis.

Disease management and therapy differ for thyrotoxicosis caused by hyperthyroidism and thyrotoxicosis caused by other conditions. Thyroid imaging and radiotracer thyroid uptake measurements, combined with serologic data, enable specific diagnosis and appropriate patient treatment.

2.1 Computer-based on education systems.

The computer-based on education systems (e.g. linear programs, graft progras, generative systems) are also known as Computer Assisted Instruction Systems (CAIS) [8]. The main deficiencies of the CAIS are:

- They try to include complete courses instead of limiting itself to concrete subjects.

- Barriers of communication between the tutor and the student, which restrict the interaction among them.

- Students do not have knowledge of how and why the tasks are executed. Also the program reacts using a series of anticipated situations independent of student's answers.

- It is not possible to transfer these systems from one domain to other. They are domain incompatible

- These systems tend to be static instead of evolving and dynamic.

- Once constructed, the knowledge that it includes is not updated with time.

In summary they are expensive and repetitive programs in which there is no relation between what is taught and how is taught. Due to these problems and efforts made by certain researchers in this area, Intelligent Tutorial Systems (ITS) are developed. The ITS combines techniques of Artificial Intelligence (AI) with psychological models of the student and the expert. It also invoves application of theories of the education [9]. We are proposing application of ITS for the education of Hyperthyroidism. In next section, we will justify the need of such ITS.

2.2 Justification

Although Hyperthyroidism is not a world-wide problem of health, the necessity to propose a CBR based system for its educational use does not arise from the number of TS incidents, but by the human aspect that chronic suffering may lead to much serious complications, if proper attention and care is not provided. Hence it would be useful to provide necessary information about the syndrome. Also CBR involves solving new problems using the solutions from the previous cases. It is a suitable approach for medical diagnosis as it involves proposing a corresponding treatment using past cases of the patients. This system is certainly valuable in the absence of an Endocrinologist or another kind of specialist at the first level of attention in various societies [10]. Medical education systems must have high quality. In the context of undertaken development, the system will involve compilation and retrospective revision of material sources i.e. clinical histories of the real cases from endocrinology medical practice [11]. In order to define strategies and content of the system, the organization and analysis of the information will require the consultation of different specialist and experts, as well as the corresponding bibliographical revision. The following examples illustrates the utility of ITS for various medical applications.

2.3 Examples of ITS

In relation to Medical Sciences, the university cannot teach all the required knowledge, facts and abilities without informing new discoveries in the form of technical advances in the medical sciences, on-line educational sources with clinical histories, images and findings in different cases from various medical institutions research about this sickness. Users can surf all the existing advances in the world of telemedicine through a virtual university. With such advances in computer science, many applications are developed for medical education [12].

A Pediatric Hospital in Germany has applied Case-based reasoning for knowledge acquisition [13]. They developed a program called "Casus" to resolve the problem of education in medicine using a case library taken from real medical practice. The program assists an apprentice to determine a diagnosis using its in-built case-library. This program

3. Development of the System

The system will justify the solution from the similar cases. The knowledge acquisition process involved at least 35 cases mainly originating from the consultations of medical experts at the first level of attention in different areas. These cases were presented to the expert so that the experts perform diagnosis, prognosis and the medical conduct. The resident specialist diagnoses these cases. The specialists are either from Endocrinologist or another kind of specialist at the first level of attention. The expert confirms the diagnosis and then only it is incorporated in the system.

3.1 Methodology

To formulate a diagnostic conclusion in Endocrinology is a complex clinical problem. To obtain the efficiency of the expert, 87% of the diagnosis given by the system must match with expert's diagnosis. Elements of observation by the expert must also be considered. Due to this complexity, there is always a scope of improvement to make the system as perfect as possible in future investigations.

Categorizing the diagnosis, prognosis and the medical conduct can test the specificity and sensibility of the system. The proposed instrument will be a medical and tool, which would be useful in orienting the medical conduct of TS. It would be more sensible and little specific to the variations in the phenomenon under study and the user's valuations. This is an important result which justifies use of such tool. Also it is guaranteed that this tool will have a capacity of giving accurate results in more than 80% of the diagnostic considerations.

The case library proposed tool would consist of use of interpretative type explanations and using archives with a similar extension [16]. It is feasible and easily transportable between different domains. It uses the option of internal correlations of the program and estimation of the weight assigned to the predicting characteristics. It defines 25 predicting characteristics and 3 objective of the system. Those are:

   i. The diagnosis in the 9 proposed groups
   ii. Directions in the edical conduct to follow
   iii. User's valuations with a total of at least 100 cases in its case library.

With this background discussion of the proposed system, we now present the development of the proposed software in the next section.

4. Developed Software

The Intelligent system for diagnosing Hyperthyroidism (SIDHYT), showed in this paper, is a program developed in Java using JBuilder 9.0 [17]. It use extension files *.shy that constitute the case library. Its application was very useful for the creation of the case-based system. It uses the option of internal correlations of the program and estimation of the weights of the predicting characteristics. It defines 25 predicting characteristics (See Figure 1) and objectives of the system that are: the diagnosis in the 9 proposed groups, directions in the medical conduct to follow and user's valuations with a total of 355 cases in its case library. The following figure shows the software interface.

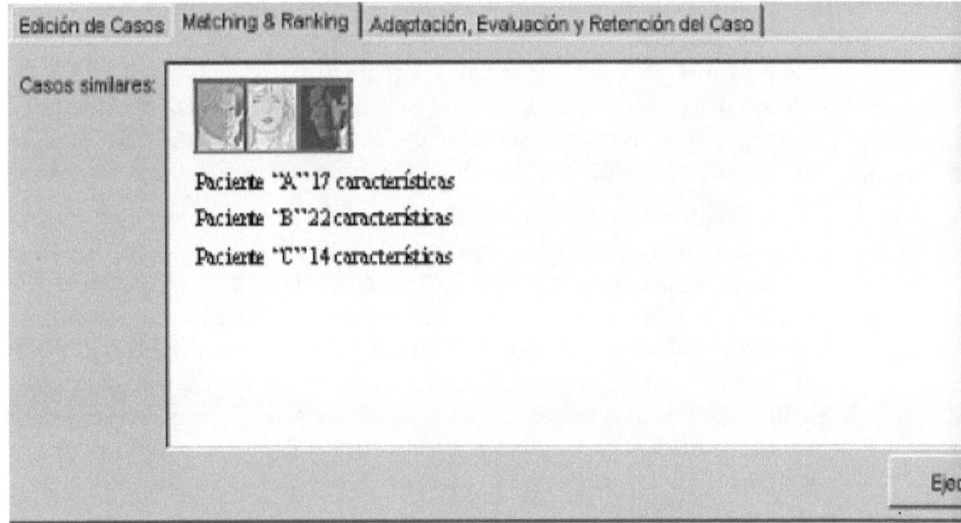

Figure 1. Developed Software Interface.

The system justifies its solution from the similar cases. It provides auto learning mechanism-using values of the predicting characteristics. It involved consultations of medical expert.

After this initial knowledge acquisition, 100 more cases unknown to the system are added. These cases were associated with psychiatric upheavals to the TS in adults.

After constructing the case-base, the software gave the following results.

- 87% of efficiency for the diagnosis
- 89% for the medical conduct to follow
- 92% for the user's valuations.

These results are considered good. It meets the aims of the medical aid. It assists the doctor in consultation. These results show the effectiveness of this software over the systems designed for study of the Hyperthyroidism at the international level. The results provided in the next section justify the usefulness of the proposed system.

5. Results

In relation to specificity sensitivity of the system, sensitivity from the instrument was 88.88%. Specificity of 85.45% with a predictive value for the true positives of 83.33% was obtained Specificity with a predictive value for the truly negative ones were 90.38%.

In relation with the treatment, sensitivity of system was 93.5%. 78.57% specificity was obtained. Specificity with a predictive value for true positives of was 91.78% and for truly negative ones was 81.48%. In case of prognosis, one obtained a sensitivity of 98.90%. Specificity of 22.22%, with a predictive value for the true positives was 92.78% and for truly negatives was 66.6% [See Table 1].

With relation to the number of cases in case library for a better expert operation, the expert with case library of 18, 36, 45, 72, 135 and 176 cases respectively was tested. The Chi-square test verified that the variables, number of cases and correct answers given by the system for the three objectives of exit had statistically significant relation for a smaller P of 0001. The tests of Spearman correlation were not statistically significant. The exact criterion about the number of optimal cases is not decided at present for a greater effectiveness of the system. With greater number of cases, the amount of correct answers from the system was increased [See Table 2].

Table 1: performance evaluation by the expert for 100 cases.

Group	VP	VN	FP	FN	S%	E%	Predictive Value VP%	Predictive Value VN%	Efficiency
Endocrinology T.	40	47	8	5	88.88	85.45	83.33	90.38	87%
Ambulatory T.	67	22	6	5	93.05	78.57	91.78	81.48	89%
Reserved Prognosis	90	2	7	1	98.90	22.22	9278	66.66	92%

Where Label:

VP: true positive.   VN: true negative.   FP: false positive.

FN: false negative.   S: sensibility.   E: specificity.

Predictive value of truly positives. Predictive value of truly negatives.

Table 2: Comparison of agreement between the experts and the human specialist for each one of objective characteristics.

Number of Cases	Agreement in diagnosis	No agreement in diagnosis	Agreement in medical conduct	No agreement in medical conduct	Agreement in Prognosis	No agreement in prognosis
18	1	17	5	13	11	9
36	8	28	18	18	29	7
45	11	34	27	18	27	18
72	26	46	63	9	69	3
135	113	22	119	16	122	13
176	154	22	157	19	162	14

Chi-square for the diagnosis = 175.641.   chi-square for medical conduct=81.669

Chi-square for the prognosis=51.735   DF=5 smaller P of 0001.

Correlation Coefficient of Sperman for diagnosis=0.11595.

Correlation Coefficient of Sperman for meical conduct = 0.34786.

Correlation Coefficient of Sperman for prognosis=0.144

Critical value -/+=0.81165 for a smaller P of 0.05

6. Conclusion

The proposal of using the case library for the education of the patient with Hyperthyroidism is based on the set of aspects that are enunciated next and that constitute results of our investigations.

1.   The results will have to at least agree with the results reported by Literature about Case-based reasoning. This tool with an acceptable

effectiveness conserves the integrity of the knowledge and provides better results It is reliable and relatively simple for its application.

2.	The proposed case library is integrated by 28 characteristics with 288 possible values or domains and will have to contain at least 100 cases. It should use the criteria of Moriyama for the validation of the content.

3.	Relation of statistically significant association with the test of Chi-square between the numbers f cases from case library exists.

4.	The created software has an effectiveness of 87% in its diagnostic considerations, 89% in his capacity of direction in medical conduct to follow and of 92% in its predicted valuations. These results are considering good in comparison with the results obtained by the other authors [18].

5.	The expert who sets out to create tool of medical aid for the diagnosis must be more specific. The expert's capacity of direction in the medical conduct to follow should be more sensible.

6.	It is tried to widely value the utility of the expert who plans to create a tool of medical aid for the boarding of the patient with Hyperthyroidism with endocrinologist upheaval.

The above results are considered good. In summary, we have proved usefulness of the CBR based system diagnosing Hyperthyroidism.

References

1.	Anderson, P., (May, 1996). Obsessive Compulsion and Tic Linked to Sore Throats. Medical the Post. http://www.mentahealth.com/mag1/fr51.html.

2.	Burden, G. (July, 1996) Imperial the Gene. Medical the Post. http://www.mentalhealth.com/mag1.

3.	Pearce, J. (1996) "Good habits and bad habits Of the life in family to the life in society", Madrid: Editions Paidós.

4.	Wagaman, J.R., Miltenberger, R.G. & Wiliams, D. et al. (1995). Treatment of to vocal tic by Differential Reinforcement. Journal of Behav. Ther. & Psychiat. 26 (1), 35-39.

5.	Bados A. (1995) "The tics and their upheavals: Nature and treatment In the childhood and adolescence". Madrid Editions Pyramid S.A.

6.	Vera, M.N. & Vila, J. (1995). Techniques of relaxation. In V. Caballo (Eds), Manual of therapy techniques: modification of the conduct, (pp: 161-181). Madrid: Veintiuno century of Spain publishing.

7.	Scotti, J.R., Schulman, D. & Hojnacki, R.M. (1994) Functional analysis and unsuccessful treatment of Hyperthyroidism in a man with mental deepest retardation. Behavior Therapy. 25, 721-738.

8. López Ostio, J. Cols, Tutorial Systems (ITS). Typed conference. San Sebastian Spain, 1993.

9. Capponi, R. (1987) Psychopathology and Psychiatry. Semiología Santiago University Editorial.

10. Polare i., Kaneshiro, Takeshi & Malashona, N. (2005) "Modelling human societies using CBR". Central Asia CCBR; Astana, kazakhstán.

11. Mink, J. W. & Weinberger, D.R., (1998). http://www.mentalhealth.com/fr20.html.

12. Mann, BD; Sarechdern, AK; Nieman, L.Z et al., (1996) "Medical Teacher to students by role playing a model integrating psychosocial issues with disease management". JI. Journal of cancer Educ, Summer 11[2]: 65-72.

13. Fisher, MR; Shaver, S; Grasel, C; Bachering, T; Handl, H; gartner, R; Scherbau, W; Scriba, PC., (1996) "Casus-Model Trial to computers-Assisted author for System Problem Oriented Learning in Medicine". Z-Artiz-Forbild-Jena, August 90 [5]: 385-389.

14. Sleeman, D.; Brown, J.S., Intelligent Tutorig Sstems, Acadeic Press London, 1982.

15. Ochoa A.; Fernandini M. & Shingareva I., (2005) "use of Oniri techniques for describing Pa'nar Syndrome". Central Asia CCBR; Astana, Kazakhstán.

16. Kolodner, JL., (1997) "Educational implications of analogy: to view from isolation cases using Case-based reasoning, Am-Psychol, Jan 31 [1]: 57-66.

17. Nieto, M.; Leguízamo-Povedano, J.; Mejía, M & Ochoa, A., (2002) "Applying dependences model to CBR software". CIIC'02; Soto La Marina, México.

18. Bichindaritz, I., (2003). Solving safety implications In a case based decision-support system in medicine. In Workshop on CBR in the Health Sciences, 9-18. ICCBR'03.

19. Bichindaritz, I., (2012): Research Themes in the Case-Based Reasoning in Health Sciences Core Literature. ICDM 2012: 9-23

Using Cultural Algorithms to determine customer behavior of a set from Casas Geo "Campo Grande" in a great city

Alberto Ochoa-Zezzatti[1], Esteban Hernandez[1], José Martínez[2] & Andrés Bautista[2]

[1]Juarez City University, México

[2]Instituto Tecnológico de Ciudad Madero

alberto.ochoa@uacj.mx

Abstract. The paper discusses a researh related with an Intelligent Recommender System which is used to improve Decision Support System based on a Hybrid Algorithm based on Social Data Mining and a Bioinspired Algorithm to determine the adequate selection of buy a house based on different features associated wth a group of seven different kind of houses, these housess conformed an "Urbanization" of 7800 Houses; we modelling different scenarios to analyze the diverse preferences of select location in this specific development and featured by socio-economical and cultural aspects, this urbanization was development in a great City in the center of Mexico, this research which permits select a specific house to live with basis on seven different kind of house, these models of houses has different size, facilities and different distance to a common swimming pool and strore of food. Each houset was analyzed to built their cost-benefit during different times and scenarios and determine the viability of buy this in the time horizon using at formal methodology based on a Bioinspired Algorithms. These houses was characterized and analyzed by obtain the most representative future scenario to determine the best location which try to improve the economical resourcesof each family and their perspectives to determine the correct selection of location. A case of study is presented regarding to the proposal horizons using data obtained from Casas Geo. The intention of the present research is to apply the computational properties; in this case understand the model of bough. In addition, we analyzed the selection and location of houses using a similarity model to locate theses. The sample of study allowed analyzing individual features of each house with the emulation from set matching features (location, price, services, in others). By means of this is possible to predict the best location to bought a house.

Keywords: Cultural Algorithms, Pattern Recognition and Decision Support System.

1.   Introduction

The concept of sustainable development is steadily approaching recognition, if not full disciplinary autonomy, becoming the focus of new theoretical and normative reflection. However, the same cannot be said of a more specific field of application of that same concept - the urban environment. In our opinion, this has been hindered until recently by some unresolved problems - of definition, methodology and epistemology - intrinsic in the more general concept, and also by some specificities of the urban case that have not been sufficiently borne in mind. This paper aims at directly facing these unresolved problems, and proposes a definition on which later empirical studies and new theoretical elaborations may be based. Urbanization is, by nature, a manufacture of people organized, an almost entirely artificial object, constructed for historical goals of socialization, synergy, increase of knowledge and social wellbeing. A "weak" concept of sustainability, which permits ample substitutability between production inputs and utility function inputs, is almost impossible to avoid. When considering the problem in its entirety, we must combine the socio-cultural, economic and environmental elements, which all go towards the construction of that complex set of relations we call an urbanization. Sustainable urban development may be defined as a process of synergetic integration and co-evolution among the great subsystems making up a city (economic, social, physical and environmental), which guarantees the local population a non-decreasing level of wellbeing in the long term, without compromising the possibilities of development of surrounding areas and contributing by this towards reducing the harmful effects of development on the biosphere.

2 Cultural Algorithms

The initial development of Cultural Algorithms (CAs) can be attributed to Reynolds [12] this approach is a complement to the metaphor used by evolutionary algorithms, which had focused on the concepts of genetics and natural evolution. Cultural algorithms are based on the theories of anthropologists, sociologists and archaeologists, who have tried to model the evolution as a process of cultural evolution [5]. The belief space characterizes CAs as evolutionary algorithms, which are used to store the acquired knowledge from previous generations. The information in this space must be accessible to any individual, who may use it to change their behavior and their respective proposed solution. To join the belief space and the population is necessary to establish a communication protocol, which dictates rules of the type of information to be exchanged between spaces. This protocol defines the acceptance and influence functions. The acceptance function is responsible for accepting the information or the experience that individuals have obtained in the current generation and transport into the belief space. On the other hand, the influence function is responsible for "influencing" variation operators (e.g. crossover and mutation in the case of genetic algorithms). This means that this function set some kind of pressure on resultant individuals from the application of variation operators to reach the desirable behavior, also away from undesirable results, always according to information stored in the belief space.

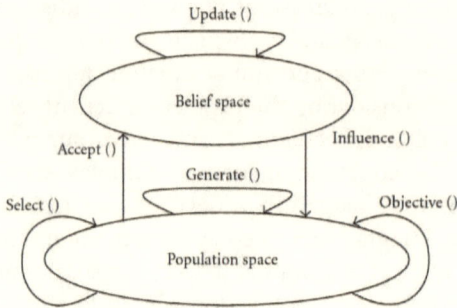

Figure. 1 Different Spaces employed by the Cultural Algorithm.

Figure 1 presents the interaction between the belief space and population space. The population space works similar to that of an evolutionary algorithm, i.e. the population consists of a set of individuals where each has an independent feature used to determine their suitability (fitness). The interaction between the two spaces makes the cultural algorithm increases the complexity in the development and computation of the evolutionary algorithm. Below we show the general pseudo-code of a cultural algorithm.

```
Begin
  t=0;
  Initialize POP(t); // Initialization of population
  Initialize BLF(t); // Initialization of believing space
  Evaluate POP(t);
  While (Do not condition of term t=t+1)
    Vote (BLF (t), Accept (POP(t))));
    Adjust (BLF (t));
    Evolve(POP(t), Influence(BLF(t)));
    t = t +1;
    Select POP(t) from POP(t-1);
  End While
End
```

3. Analysis of Customer Behavior of bought related with an Urbanization in a great city.

We try to improve the next equation 1 using the table 1, in which is V as the value of the property and define with different aspects relative with the potential features of the house.

$$V = \left\{ \frac{CHU}{IHI} \left[\frac{\#R * SH}{PLH} \right]^{RA} \pm SCAH \right.$$

In where:

CHU = Cost of House Unit.

IHI = Increase House Index

#R = Number of rooms in the house.

SH = Size of House.

PLA = People living on the house.

RA = Recreational Areas.

SCAH = Social Cultural Aspects related to the location of the House (Space to parking, access to schools, security and others).

Table 1. Multivariable analysis with the information related with each house and its components.

Model of House	House Increase Index	Commercialization	Services	Recreation and green areas	Cost-Benefit	Equipment in the House
Barbados	8	0.814	0.765	0.863	0.799	0.678
Crimea	14	0.795	0.811	0.835	0.847	0.715
Djibouti	5	0.747	0.838	0.842	0.817	0.818
Macao	8	0.816	0.794	0.803	0.858	0.902
Pitcairn	9	0.947	0.836	0.828	0.807	0.794
Tahití	12	0.877	0.819	0.842	0.805	0.816
Tuvalu	7	0.954	0.797	0.783	0.912	0.828

4 Multiple Matching

The multiple matching is a series of seven evaluations according to different combinations of houses and a batch of 50 runs under different scenarios. In the evaluation phase economics specifications with more similarities will be given a preference, and then these aspects will be selected to compete. Each house makes a compromise and participates in exactly seven of these evaluations. Houses must be ranked according to their customers' preferences after tournaments end once the final list of multiple matching is evaluated. The hybrid algorithm sets the right for customers to evaluate a batch according to the organizational needs and the houses for each comparison assign the houses list before a new cycle begins. Each evaluation will have all the houses playing over a schedule of seventeen runs.

The hybrid algorithm will be scheduled to set the timing for the comparison of different similarities using a round of multiple matching analyses based in the commercialization assigned to a house. Then, houses that qualify for selection in a Model will be chosen on the following prioritized basis.

For the first cycle of similarity, all houses in the Repository (ie. Barbados or Tahití houses model) will be invited to participate for different comparisons. Given the organization for each house and the matches for each round in the algorithm, houses are asked to state their participation for its evaluation in each of the series. In case any of these house decline to participate in the series, the algorithm may nominate one house to be set as a replacement, and this house has to be rated amongst the top houses in the Repository. Based on an average calculation of two decimal places, the rating list in the series of comparisons, before starting a new cycle, three qualifiers will be selected (excluding the seven houses that will be compared in the matches). In case houses have the same average rating, the number of similarities set for the match will be used to determine its ranking.

To ensure an active participation in the future, a minimum of twenty-five games are recommended for the four included rating lists and before the main rating list. When a house does not accept to play into a Multiple Matching series, then the selection process uses the average rating plus number of games played during the rating period. The algorithm repeats this process until reaching the required qualifiers of the Multiple Matching series and location to each house and the real possibility of bought.

5 Experimentation

In order to obtain the most efficient arrangement of houses, we developed a cluster for storing the data of each of the representative individuals for each house. The narrative guide is made with the purpose of distributing an optimal form for each the evaluated houses [9]. The main experiment consisted in implementing houses in the Cultural Algorithm, with 500 agents and 250 beliefs into the belief space. The stop condition is reached after 75 runs; this allowed generating the best selection of each kind and their possible location in a specific Model. A location is obtained after comparing the different cultural and economical similarities of each house and the evaluation of the Multiple Matching Model as in [10]. The vector of weights employed for the fitness function is $W_i=[0.6, 0.7, 0.8, 0.5, 0.6, 0.7]$, which respectively represents the importance of the particular attributes: House Increase Index, Commercialization, Services, Recreation and green areas, Cost-Benefit and Equipments on the house . Then, the cultural algorithm will select the specific location of each house based on the attributes similarity. Each attribute is represented by a discrete value from 0 to 7, where 0 means absence and 7 the highest value of the attribute. The experiment design consists of an orthogonal array test with interactions amongst the attribute variables; these variables are studied within a location range (1 to 400) specific to a coordinates x and y. The orthogonal array is L-N(2**5), in other words, 6 times the N executions. The value of N is defined by the combination of the 6 possible values of the variables, also the values in the location range. In Table 3 we list some possible scenarios as the result of combining the values of the attributes and the specific location to represent a specific issue (house). The results permit us to analyze the effect of the variables in the location selection of all the possible combinations of values.

Table 3 The orthogonal array test.

House Increase Indexs	Commercializati on	Service	Recreation and Areas	Cost- GreenBenefint	Equipme on the House
4	1	2	2	3	4
3	1	2	2	3	3
2	1	3	2	4	4
5	1	3	2	5	2

The use of the orthogonal array test facilitates the reorganization of the different attributes. Also the array aids to specify the best possibilities to adequate correct solutions (locations) for each house. Different attributes were used to identify the real possibilities of improving a house set in a particular environment, and to specify the correlations with other houses (see Figure 2). The locations will be chose based on the orthogonal test array

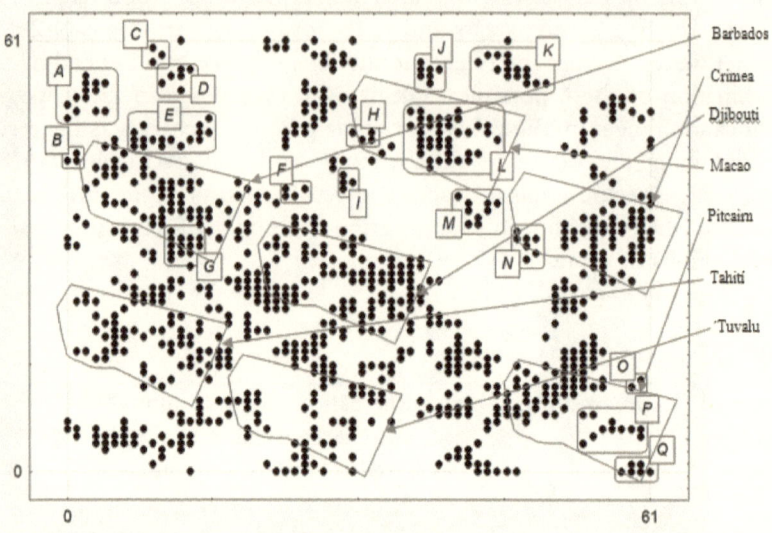

Figure 2.- Location of houses bought in the urbanization. The cluster B, C, D, F, H, I, M, N, O, and Q are more related with people in families (brothers, sisters, uncles, aunts), the clusters A, E, G, J, K, L and P are clusters more related with friendship relationships.

6 Conclusions

After our experiments we were able to remark the importance of the diversity of the established economical patterns for each house. These patterns represent a unique form of adaptive behavior that solves a computational problem that does not make clusters of the houses.

The resultant configurations can be metaphorically related to the knowledge of the behavior of the community with respect to an optimization problem (to culturally select 5 similar houses [4]). Our implementation related each of the houses to a specific a location quadrant. The Narrative guide, allowed us to identify changes in time related to one or another house. Here, we show that the use of cultural algorithms substantially increased the understanding in obtaining the "best paradigm". This after the classification of agent communities was made based on a relation that keeps their attributes. Therefore, we realize that the concept of "negotiation" exists based on determining the acceptance function to propose an alternative location for the rest of the houses [8]. For further implementations we intend to analyze the level and degree of cognitive knowledge for each house. Additionally, this may help to understand true similarities that share different houses based in the characteristics to be clustered and also to keep their own biological identity. In a related work [7], it has been demonstrated that small variations go beyond phenotypic characteristics and are mainly associate to tastes and related characteristics developed through the time. On the other hand, CAs can be used in the Evolutionary Robotic field where social interaction and decision is needed, for example in the training phase described in [11], and to organize group of robots for collaborative tasks. Another future work using CAs is related to the distribution of workgroups, social groups or social networking. Finally, CAs can be used in pattern recognition in a social database, for example: fashion styling and criminal behavior.

Acknowledgements

We thank the community of Computational Optimization that for a decade has researched about novel ways to display social topics.

References

1. Desmond, A. and Moore J. (1995) Darwin – la vida de un evolucionista atormentado. Generación Editorial, São Paulo, Brazil
2. Ochoa A. et al. (2007) Baharastar – Simulador de Algoritmos Culturales para la Minería de Datos Social. In Proceedings of COMCEV'2007
3. Ochoa, Alberto et al. (2009) Dyoram's Representation Using a Mosaic Image, The International Journal of Virtual Reality
4. Memory Alpha, http://en.memory-alpha.org
5. Callogerodóttir Z. and Ochoa A. (2007) Optimization Problem Solving using Predator/Prey Games and Cultural Algorithms NDAM'2003, Reykiavik; Iceland
6. Tang Hué et al. (2006) The Emergence of Social Network Hierarchy Using Cultural Algorithms, VLDB'06, Seoul, Korea
7. Vukčević I. and Ochoa A. (2005) Similar cultural relationships in Montenegro JASSS'2005, England
8. Zuckermann Dennis (1991) Culture and Organizations, London: McGraw-Hill,
9. Ponce J. et al. (2009) Data Mining and Knowledge Discovery in Real Life Applications", Book edited by: Julio Ponce and Adem Karahoca, ISBN 978-3-902613-53-0, pp. 438, I-Tech, Vienna, Austria
10. Ustaoglu, Yesim. (2009) Simulating the behavior of a minority in Turkish Society ASNA'2009, Zurigo, Svizerra
11. Nolfi S. and Floreano D. (2000) Evolutionary Robotic: The Biology, Intelligence, and Technology of Self-Organization Machines, MA: MIT Press.

12.   Robert G. Reynolds and William Sverdlik. (1994) Problem Solving Using Cultural Algorithms. International Conference on Evolutionary Computation, 645-650

Applying a Bioinspired Algorithm to improve a Businessgames based on strategic planning

Alberto Ochoa-Zezzatti, [1], Guadalupe Gútierrez[2], Lourdes Margain[2] & Rubén Jaramillo[3]

[1] Juarez City University, México

[2] Universidad Politécnica de Aguascalientes, México

[3] CFE-LAPEM, México

alberto.ochoa@uacj.mx

Abstract. In the present study shows the design of a tool based on bio-inspired algorithm as an aid in the strategic planning methodology, unlike current applications, is focused on an environment that goes beyond simple numerical forecasts and statistical processes. It is based on an advanced algorithm that optimizes the strategies to be followed within the company, helping businesses to achieve competitive advantage in the market. This model is flexible, adaptive, has learning ability, is robust and fault tolerant. This methodology provides optimal strategies to improve competitiveness of a company, the ability of the model can provide strategies that are not obvious because they can find no obvious relationships among variables can help the manager or leader of an organization. This tool is an aid in the process of improving competitiveness because it supports the strategic decisions made in administrative levels of the organizations.

Keywords: Strategic Planning, Decision Support System, Bioinspired algorithm.

Introduction

There are significant benefits to using a clear process of formulating a strategy in pursuit of the realization of the goals set for a company, this in order to ensure that at least the functional departments policies are coordinated and directed a group objectives (Porter, 1998). The company strategy is the action plan of management to operate the business and manage its operations. That is, it's how to play a strategic vision and achieve objectives. Likewise, efforts to plan and execute a strategy are vital administrative functions and excellent execution of an excellent strategy is the most reliable recipe for turning a company into one of the best of its kind (Thompson et al, 2008). Through good strategic planning a company can achieve great results as well as being of vital importance to address the changes and challenges they face in business in a highly competitive global market. There are already several models for strategic planning, such as the five generic competitive strategies mentioned the book Strategic Management Thomson. However, companies still face the problem of what to use or what strategy is best for the long and short term. The purpose of this study is to design a tool based on a bioinspired algorithm to support strategic planning methodology, based on an advanced algorithm that optimizes the strategies to be followed within the company, helping businesses to achieve competitive advantage in the market and identify opportunities. This model is flexible, adaptive, has learning ability, is robust and fault tolerant. All these features allow mentioned over time improving strategies or redefining go efficiently. Flexibility and adaptability are important factors because now the ability to react quickly to unexpected changes in the market can make the difference between success or bankrupt a company. The model is robust and fault tolerant, since this feature is typical of models based on evolutionary computation. Overall, this tool allows the continuous improvement of the strategies used to achieve success in a company because the system will be able, through an advanced algorithm, to identify opportunities and / or threats facing the corporation and assess what Strategies can fortify the course of action of the corporation to find or improve competitive advantage, enabling the management or administrative leader decision making. **The present research is organized as follows. In the literature review section presents relevant literature on the background in the process of decision making and strategic planning bioinspired about this algorithm. In the methodology section presents the whole design of the strategic planning tool based on evolutionary computation. Finally we present the conclusions of this research and the limitations of the study and recommendations for future research.**

2. Bioinspired Algorithm

Firefly Algorithm

The Algorithm "Firefly algorithm" (FA) is a metaheuristic algortmo, inspired by the comportment of tildeo of fireflies. The primary purpose of the flash of a lúciernagapara The primary propossito for flashing of a firefly is to act coo or signal system to attract other fireflies. Xin-She Yang firefly algorithm makes the following criteria:

1. All the fireflies fireflies are unisexual, so any firefly can be attracted by either of the other fireflies;

2. Its appeal is proportional to the degree of brilliance, and for any two fireflies, the dimmest of them may be attracted by (and thus move toward) the brightest of them, however, the brightness may decrease as the distance is increased;

3. If there are no fireflies firefly brighter than a given, this will move randomly.

The brightness may be associated with the objective function ..

Firefly Algorithm or metaheuristic optimization algorithm inspired by nature.

1) Objective function: $f(\mathbf{x}), \quad \mathbf{x} = (x_1, x_2, ..., x_d)$;

2) Generate an initial population of fireflies $\mathbf{x}_i \quad (i = 1, 2, \dots, n)$;.

3) Formulate light intensity I so that it is associated with $f(\mathbf{x})$

(for example, for maximization problems, $I \propto f(\mathbf{x})$ or simply $I = f(\mathbf{x})$;

4) Define absorption coefficient γ

While (t<MaxGeneration)

  for i=1:n (all n fireflies)

    for j=1:n (n fireflies)

     if ($I_j > I_i$),

      move firefly i towards j;

     end if

     Vary attractiveness with distance r via $\exp(-\gamma\, r)$;

     Evaluate new solutions and update light intensity;

    end for j

  end for i

  Rank fireflies and find the current best;

end while

Post-processing the results and visualization;

end

The main formula update for any pair of two dragonflies are $\mathbf{x}_i$ and $\mathbf{x}_j$ is

$$\mathbf{x}_i^{t+1} = \mathbf{x}_i^t + \beta \exp[-\gamma r_{ij}^2] + \alpha_t \epsilon_t$$

where α_t is a parameter controlling the step size, while ϵ_t is a vector is drawn from a Gaussian distribution or otherwise.

This can be demonstrated through the limiting case $\gamma \to 0$,the particle which corresponds to the standard PSO (Particle Swarm Optimization (PSO). In fact, if the inner loop (for j) is removed and the brightness I_j is replaced by the current global g^*, wholesale then essentially FA is becoming the standard PSO .

Implementation Guide

The term γ may be related to the scales of design variables. Ideally, the term β may be ordered, which requires the γ may be linked by ladders. For example, one possible choice is to use $\gamma = 1/\sqrt{L}$ where L is the average scale of the problem characterized. In the event that the scales vary significantly, γ can be considered as a vector to accommodate different levels in different dimensions. Similarly, α_t may also be linked with scales. For example in $\alpha_t \leftarrow 0.01 L \alpha_t$. It is worth noting that the above description does not include reduction of randomness. In fact, in the actual execution by most researchers, the movement of the firefly is gradually reduced by a reduction of the randomness of the type "annealing" as through

$\alpha = \alpha_0 \delta^t$ where $0 < \delta < 1 (e.g. \delta = 0.97)$. In a problem with a high degree of difficulty, can be useful if some stages is increased, then reduce it when necessary. This nonmonotonic variation of the algorithm will escape any local optimum when in the unlikely event that could get stuck if the chance is reduced too quickly. Parametric studies show that n (number of fireflies) should be 15 to 40 for most problems. An implementation on Python language is also available, although with limited functionality. Recent studies show that the algorithm is very efficient firefly, [11], and may outperform other metaheuristics based algorithms including particle swarm optimization (PSO). Most metaheuristic algorithms may have difficulty dealing with stochastic test functions, and it looks like fireflies algorithm can deal with stochastic test functions very efficiently. In addition, the FA algorithm is also better to deal with noisy optimization problems with the ease of implementation. Chatterjee et al. [12] shows that the algorithm exceeds firefly particle swarm optimization in some applications. Furthermore, the algorithm of fireflies can efficiently solve problems with non-convex complex nonlinear constraints. Further improvements in performance are possible with promising results.

3. Literture Review.

There are significant benefits to using a clear process of formulating a strategy in pursuit of the realization of the goals set for a company, this in order to ensure that at least the functional departments policies are coordinated and directed a group objectives (Porter, 1998). The company strategy is the action plan of management to operate the business and manage its operations. That is, it's how to play a strategic vision and achieve objectives. Likewise, efforts to plan and execute a strategy are vital administrative functions and excellent execution of an excellent strategy is the most reliable recipe for turning a company into one of the best of its kind (Thompson et al, 2008). Through good strategic planning a company can achieve great results as well as being of vital importance to address the changes and challenges they face in business in a highly competitive global market. There are already several models for strategic planning, such as the five generic competitive strategies mentioned the book Strategic Management Thomson. However, companies still face the problem of what to use or what strategy is best for the long and short term. The purpose of this study is to design a tool based on a bioinspired algorithm to support strategic planning methodology, based on an advanced algorithm that optimizes the strategies to be followed within the company, helping businesses to achieve competitive advantage in the market and identify opportunities. This model is flexible, adaptive, has learning ability, is robust and fault tolerant. All these features allow mentioned over time improving strategies or redefining go efficiently. Flexibility and adaptability are important factors because now the ability to react quickly to unexpected changes in the market can make the difference between success or bankrupt a company. The model is robust and fault tolerant, since this feature is typical of models based on evolutionary computation. Overall, this tool allows the continuous improvement of the strategies used to achieve success in a company because the system will be able, through an advanced algorithm, to identify opportunities and / or threats facing the corporation and assess what Strategies can fortify the course of action of the corporation to find or improve competitive advantage, enabling the management or administrative leader decision making. The present research is organized as follows. In the literature review section presents relevant literature on the background in the process of decision making and strategic planning based on this bioinspired algorithm. In the methodology section presents the whole design of the strategic planning tool based on evolutionary computation. Finally we present the conclusions of this research and the limitations of the study and recommendations for future research.

3.   Modelling.

A bioinspired algorithm from the perspective of strategic planning is a tool for decision making, which is used as a support by the administrative or executive leaders who are responsible for making business decisions, in order to select a response appropriate for a particular situation. The development of this tool consists of a series of stages that must be examined step by step. Beginning with the definition of variables which provide the input information to the system and to be processed by the same, followed bioinspirdo design algorithm which is responsible for providing the artificial intelligence of this tool, continuing the definition of outputs which provide support decision making and finishing processes with use, application and system training. This decision support system designed to help the administrator in the process of implementing the strategies, not during the formulation process, but rather to help monitor and improve the strategies that have been formulated and are being implemented in a . The system will prompt administrative leaders can use complementary strategy to strengthen its current strategy. Complementary strategies are actions undertaken to achieve administrators chosen strategy, these strategies are chosen based on their applicability to specific market conditions (Petrorius, 2008). Also, the system will be able, through an advanced algorithm, to identify opportunities and / or threats facing the corporation and assess which strategies can strengthen the course of action of the corporation to find or improve competitive advantage. Input information for decision support tool for the design of this tool the first element that is taken into account is the information that the system requires, since this information will apply intelligent algorithm based on evolutionary computation and produce a output that supports strategic decisions. The information required as input to the system is divided into two groups, the first is the current status information of the company and the second is the future state which is expected to arrive. Table 1 shows the system input variables corresponding to the present state of the company and the nomenclature to be used in the design of the tool. An important feature of the group of variables is the type that are quantitative.

Table 2 presents the input variables of the system for the future state which is expected to hit the company. These values serve as reference for the tool to assess which strategy is the best option to achieve the desired objectives. After identifying all groups of variables is necessary to construct the vector of the input data to the tool, as expressed in the equation 1.

$$X = \begin{pmatrix} B_1, B_2, \dots, B_n, R_1, R_2, \dots, R_n, P_1, P_2, \dots, P_n, F_1, F_2, \dots, F_n, E_1, E_2, \dots, E_n, \\ C_1, C_2, \dots, C_n, Y_1, Y_2, \dots, Y_n, D_1, D_2, \dots, D_n, M_1, M_2, \dots, M_n \end{pmatrix} \quad (1)$$

4   Design of experiments

If a dispute arises between the output certainty under the proposed algorithm and the desired output, the gradient descent algorithm is used to adjust the solution within the search space. Otherwise, there is a viable solution. This process represents the steps that a manager to make decisions at the stage of finding the best solution, the leader receives all necessary information about the internal and external situation of the company, equivalent to the power of the algorithm; once all information has performed an internal process of reasoning and evaluation of possibilities, processing equivalent to performing the comparative particle algorithm, then makes a decision based on your reasoning, equivalent to the output of the algorithm. In the evaluation phase of the algorithm the manager evaluates the decision, then analyzes its possible implications and whether there is any error or change its strategy set. But the manager would not have made this learning process does not exist.

Table 1: Input Variables Present State Company

Group variables	Nomenclature	Description
Evaluation of the current strategy of the company	B1, B2, ... , Bn	This set of variables contains all the information corresponding to the indicators that measure the performance of the current strategy. An example of these metric variables are contained in the balanced scorecard (Balanced Scorecard).
Identification of strengths, weaknesses, opportunities and threats	R1, R2,...,Rn	This group of variables containing the evaluation of the strengths and weaknesses of resources of a company, its business opportunities and threats outside of their future welfare. An example of this group of variables is the SWOT analysis indicators, financial ratios, among others.
Assessment rates and	P1,P2,...,Pn	This set of variables contains all

business costs.		information relevant to the price and current costs of the company. For example internal margin, the margin suppliers, value chain, benchmarking, in another..
Competitive strengths	F1,F2,…,Fn	This set of variables contains a group of measurements of strengths competitive, such as quality, reputation, production capacity, capacity technology, distribution network, in another
Dominant economic characteristics of the industry.	E1,E2,…,En	This group of variables contains the dominant economic characteristics industry such as market size, growth rate, number of ivals, degree of product differentiation, rate of technological change, economies of scale, in anther..
Evaluation of competitive forces	C1,C2,…,Cn	This group is composed of variables competitive forces facing industry and the intensity of each. Within this group you can find the evaluation of substitutes, degree of influence of buyers and suppliers, potential new players, among others.

This table shows the input variables that represent the current state of the company, which will feed the decision support tool.

Table 2: Input Variables Future State Enterprise

Group variables	Nomenclature	Description
Budgets and projections	Y1,Y2,…,Yn	This group of variables contains all projections, budgets and objectives raised in the strategy
Variables that measure change and its effects.	D1,D2,…,Dn	This group consists of variables that measure the effects of changes to the that the company may face in the market. Within this group can be find the rate of growth, technological change and innovation in the manufacturing, marketing innovation, and regulatory influence changes in

		government policies, changes in the concerns, attitudes and lifestyles in a society, among others.
Market position and competitors	M1,M2,...,Mn	This set of variables contains all the parameters evaluating the position which engages in the market, as well as the position of the competitors

This table shows the input variables that represent the current state of the company, which will feed the decision support tool.

5   Conclusions.

In conclusion, through the design of this tool was found that the use of Evolutionary Computation provides an important support in the strategic decisions are made in a company. In this paper we provide a design with the attributes of flexibility, adaptability and learning ability, as their previous training allows the tool to be adapted to any business environment, this represents an output suitable for the type of complementary strategy that can be taken to reinforce the main strategy and find the hits and misses that the company experienced. In addition, the design is fault tolerance, this means that if there is an abnormality in any of the variables of the input information, the tool will have a margin of error that will not allow the output is affected. In the human brain, when faced with a new problem, a solution our brain recovers from the past and adapt and allow the brain continue to function, it very closely describes the fault tolerance of the tool and the process that we would occurred within a company on the preservation of knowledge. Consequently, artificial neural networks are useful in environments that go beyond simple numerical forecasts and statistical processes. The design of the proposed strategic planning tool internally makes use of such forecasts and statistical processes, but also diagnosis, and strategy solutions to problems taking into account the competitive environment in which the company is located and identification of strengths and weaknesses same, giving an extra dimension necessary for increasingly dynamic and unpredictable business environments. This tool does not replace the manager or administrative leader responsible for making decisions but helps potentiate their skills helping to find non-obvious relationships between variables, in addition to preserving the knowledge generated in the company, making the process of learning a new manager or leader. All this with the aim of, through the optimal use of the capabilities of management, to achieve a highly competitive position in the company. Finally, future research plans to bring the theoretical development of this tool to a phase of implementation and testing. Were selected academic and business environments for testing by a group of experts in strategic management, which can measure both qualitatively and quantitatively the benefits of proposed based strategic tool and evaluate performance.

References

13. Barnes, J. (1984). Cognitive Biases and Their Impact on Strategic Planning. Strategic Management Journal, 5(2), 129.

14. Kanooni, A. (2009). Organizational factors affecting business and information technology alignment: A structural equation modeling analysis. Ph.D. dissertation, Capella University, United States -- Minnesota.

15. Klayman, J. , & Schoemaker, P. (1993). Thinking about the future: A cognitive perspective. Journal of Forecasting, 12(2), 161.

16. Laudon K. & Laudon J. (2002). Management Information Systems Mamaging the Digital Firm. (7ª Ed). E.U.A. Pearson Prentice-Hall. (pp 401- 465).

17. Peyrefitte J., Golden P., & Brice J. Jr. (2002). Vertical integration and economic performance: A managerial capability framework. Management Decision, 40(3), 217-226.

18. Philip, G. (2007). IS Strategic Planning for Operational Efficiency. Information Systems Management, 24(3), 247-264.

19. Porter, M. (1998). Estrategia Competitiva: Técnicas para el Análisis de los Sectores Industriales y de la Competencia. México: Continental.

20. Pretorius, M. (2008). When Porter's generic strategies are not enough: complementary strategies for turnaround situations. The Journal of Business Strategy, 29(6), 19-28.

21. Salem, M. (2005). The Use of Strategic Planning Tools and Techniques in Saudi Arabia: An Empirical study. International Journal of Management, 22(3), 376-395,507.

22. Sundin S., & Braban-Ledoux C. (2001). Artificial Intelligence–Based Decision Support Technologies in Pavement Management. Computer-Aided Civil & Infrastructure Engineering, 16(2), 143.

Ubiquitous Mobile to prevent Carjacking

Alberto Ochoa[1], Rubén Jaramillo[2], Arturo Hernández[3], Iván Soto[1] and Nemesio Castillo[1]

[1] Juarez City University, México

[2] CFE-LAPEM.

[3] CIMAT Research Center, México.

alberto.ochoa@uacj.mx

Abstract Juarez City, is the fourth largest populous city in Mexico located in the border with United States over a million of people in less of 75 km^2 with many problems associated with violence and insecurity provoked by organized crime, such as assaults, kidnappings, multi homicides, burglary, among other consequences. The population of the city to take action of different kinds to minimize carjacking issues related with 27857 attacks during this period of time; the reason which we present the following project, which aims to provide at people which drive over city with a technological tool that provides an indicator to the user information based statistics compiled by the Centre for Social Research at Juarez City University and public sources so uncertain is the place where you drive according a time and date specific. This research try to combine a Mobile Device based on Cultural Algorithms and Data Mining to determine the danger of suffer an attack of carjacking in a part of the city during a specific time.

Keywords: Data Mining, Cultural Algorithms and Mobile Devices.

Introduction

Carjacking is a form of hijacking, where the crime is of stealing a motor vehicle and also armed assault when the vehicle is occupied. Historically, such as in the rash of semi-trailer truck hijackings during the 1960s, the general term hijacking was used for that type of vehicle abduction, which did not often include kidnapping of the driver, and concentrated on the theft of the load, rather than the vehicle itself. During the later day car theft crime, typically, the carjacker is armed, and the driver is forced out of the car with the threat of bodily injury. In other rarer cases, the driver is kidnapped under the assault by a weapon and is retained as a passenger under duress, or made to drive his or her abductor. Women are particularly victimized in this latter method for this reason we try to achieve this problem, using data analysis. All of them joining with the aim of this project consist in a mobile geographic information system

93

(SIGMA) on levels of carjacking in areas of Ciudad Juarez, which is presented in detail below, this research include the analysis of Cultural Algorithms to determine the danger of suffers carjacking during a specific time. We realize an exhaustive analysis of other similar research, the only similar context is explain in [6], where is calculated the insecurity associate with suffering an attack of carjacking to a vehicular group which requires delivery products in different places with randomly scheduling, but this research don't considering real statistics on time and the perspective to suggest a different route to travel or stay during many time.

Project Development

To do this research project was developed by dividing into three sections which are modules of application development, implementation of the server and the intelligent module associated with Cultural Algorithms and Data Mining. Android is the operating system that is growing in to Streak 5 from Dell, for this reason we select this mobile dispositive along with other manufacturers are propelling the Latin American landing on Android with inexpensive equipment, and on the other hand, some complain about the fragmentation of the platform due to the different versions. Android is free software, so any developer can download the SDK (development kit) that contains your API [2]. This research tries to prevent carjacking attacks in Juarez City where the deaths by 100,000 people are 27 during 2011.

Components of the application

The first step is to get the coordinates where the user is located, and later sent to the server when it receives and calculates the number of carjacking within the radio closest specified in the configuration of the mobile device, the results are processed to determine a numerical index which will then be represented by a color for better visualization of the user, all this information is obtained from databases and analyzed with Data Mining. After the values are sent to the mobile device and interpreted to construct a URL to be sent to Google Maps API with which to get a map with the indicators in the area. The figure 1 shown below represents generally the operation of the Sigma, which is divided into two parts, the mobile application and Web services hosted on the server. The database in which data is stored on carjacking occurred in Ciudad Juarez, which has 3 tables:

Type_carjacking: In which are characterized the types of carjacking defined.

Neighborhood: Here is the georeference of the neighrborhood.

Attacks: It lists the incidents raised in the different regions of Juarez City, which are also expressed in decimal degrees georeferenced to facilitate further calculations, and a narrative guide associated with features of the attack.

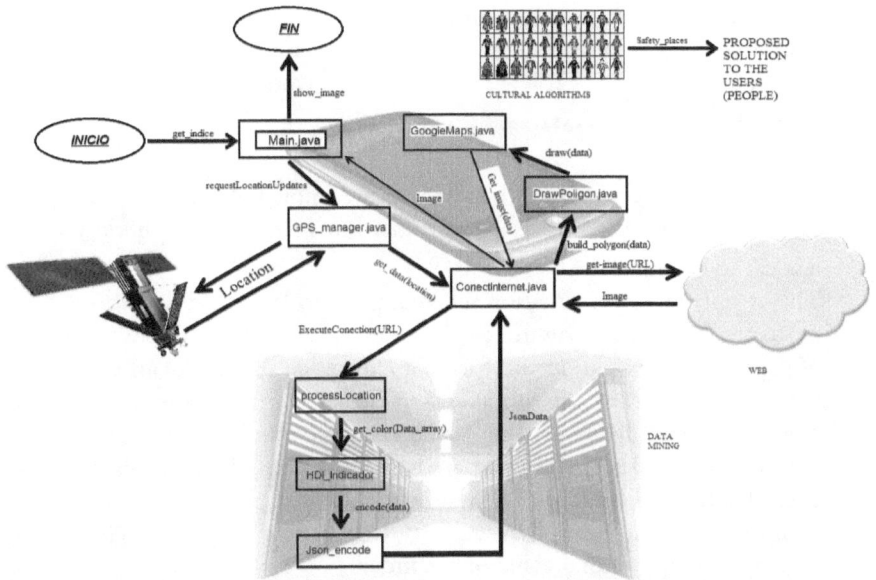

Fig. 1. Functional diagram of the SIGM including hybrid methodology.

The structure of the database has a table called attacks, where each record contains the geographic coordinates expressed in decimal degrees, which determine specific position of the attacks [1] and is visualized I Figure 2. To locate nearby points within a radius requires four parameters: latitude, longitude, and distance equatorial radius, the latter determines the maximum distance the search radius, these parameters are part of a method based on Haversine formula, which is used to calculate great circle distances between two points on a sphere. Haversine function that is different from Haversine formula is given by the function of semiverseno where:

$$semiversin(\theta) = haversin\,(\theta) = \frac{versin(\theta)}{2} = sin^2\left(\frac{\theta}{2}\right)$$

Haversine formula for any two points on a sphere:

d is the distance between two points (along a great circle of the sphere),
R is the radius of the sphere,
φ 1 is the latitude of point 1,
φ 2 is the latitude of point 2, and
$\Delta\,\lambda$ is the length difference

$$haversine\left(\frac{d}{R}\right) = haversin\,(\varphi1 - \varphi2) + cos(\varphi1)\,cos(\varphi2)haversine(\Delta\,\lambda)$$

Solving Haversine formula can calculate the distance, either by applying the function inverse Haversine or by using the arcsine where

$$h = haversine\left(\frac{h}{R}\right)$$

95

$d = R\ haversine^{-1}(h) = 2R\ arcsin(\sqrt{h})$

With this formula one can construct the following instructions, taking into account the role of semiverseno we have:

$$h = \sin^2\left(\frac{\varphi 1 - \varphi 2}{2}\right) + \cos(\varphi 1).\cos(\varphi 2).\sin^2\left(\frac{\Delta\lambda}{2}\right)$$

$$d = 2R\ arcsin(\sqrt{h})$$

For performance reasons when Haversine implement the function using SQL statements used a similar formula but in terms of spherical cosine law called cosine $\cos(c) = \cos(a)\cos(b) + \sin(a)\sin(b)\cos(C)$

This is only an approximation when applied to land, and that this is not a perfect sphere radius of the earth varies as we approach the poles. The SQL statement based on the formula Haversine find all locations that are within the range of the variable \$ radius and groups them by type of carjacking. Calculate of the uncertainty is modeled on the human development index (HDI) prepared by the United Nations Program for Development (PNDU). The insecurity index is composed of two components Number of Attacks (Q) and time (H) each component represents half the total value of the index:

Each expressed with a value from 0 to 1 for which we use the following general formula:

Component index = (Value – minimum) / (Maximum – minimum)

The component Q is calculated by the number of attacks in the area, determined by the position to calculate the maximum and minimum, accidents are grouped by neighborhood, according to the CIS-ICSA at Juarez City University is near to "Cuatro Siglos Avenue" with 35 attacks, which is the maximum, then an avenue could exceed this number and then would be the new high which means that the maximum is dynamic depending on the data entering the database. To calculate the H component data are grouped in ranges of time, for example suppose that areas of 300 m radius 13 deaths are grouped in the following form and consult the index at 4:38 pm, as is shown in Table 1.

Table 1 Example of Neighborhood number of homicides range by hour.

Amount	Time Range
5	5:00 a 6:00 pm.
3	4:00 a 5:00pm.
3	11:00 a 12:00pm.
2	9:00 a 10:00am.

The time when we consulted put us in the range of 4:00 to 5:00 pm. And the calculation of the H component would be: $H = \dfrac{3-2}{5-2} = .333$

Using data from the previous example would be the component Q:

$Q = \dfrac{13 - 2}{53 - 2} = .215$

The index would be as follows $indice = \frac{1}{2}(215) + \frac{1}{2}(333) = .274$

Having calculated the numerical index is assigned a color depending on the range in which is positioned this is very important to prevent a carjacking attack, according to the table 2 divided into eight classes.

Table 2 Ranges of colors according to the numeric index.

0 - .125		.56 - .625	
.126 - .25		.626 - .75	
.26 - .375		.751 - .875	
.376 - .5		.876 - 1	

Fig. 2. Intelligent Tool recommend another two scape routes after obtain data from a carjackig.

Implementation of the Intelligent Application

When designing an interface for mobile devices has to take into account that the space is very small screen, plus there are many resolutions and screen sizes so it is necessary to design an interface that suits most devices. This module explains how to work with different layout provided by the Android API. The programming interface is through XML. Obtaining the geographical position of a device can be made by different suppliers; the most commonly used in this project through GPS and using access points (Wi-Fi) nearby, and perform the same action but differ in some as accuracy, speed and resource consumption. Data Server Communication is the module most important because it allows communication with the server, allowing you to send the GPS position obtained by receiving the processed image and map of our location, thus showing the outcome of your application that is the indicator of insecurity. To communicate to a server requires a HTTP client which can send parameters and to establish a connection using TCP / IP, client for HTTP, can access to any server or service as this is able to get response from the server and interpreted by a stream of data. The Android SDK has two classes with which we can achieve this, HttpClient and HttpPost. With the class HttpClient is done to connect to a remote server, it needs HttpPost class will have the URI or URL of the remote server. This method receives a URL as a parameter and using classes HttpPost HttpClient and the result is obtained and received from the server, in this specific case is only text, which can be JSON or XML format. Here, the server responds with a JSON object which would give the indicator is then used to create the map. For the construction of the polygon that indicates the rate of incidents in a certain radius of the current position is not possible to create it using the GPS coordinates that yields, as these are specified in "degrees" and requires the unit to convert to meters. For this you need to know how an arc equals the terrestrial sphere, which depends on the place on earth where it is located and the address where you are, the simplest case is to measure an arc in Equator, considering that the earth is 3670 km radius, the perimeter of serious Equator radio 2, which would be equal to 40.024 miles. With this you can get a relationship that would be as follows. If 360 degrees is 40.024 miles then a degree is 111,000.18 miles, this relationship can add and subtract yards to the position, as shown in the figure 3.

Fig. 3. Acquisition of polygon map with the position of the Android application.

For the preparation of graphics, we propose use a class supported with Cultural Algorithm supported which facilitates the manipulation of data to express visually using different types of graphs, as in the figure 4.

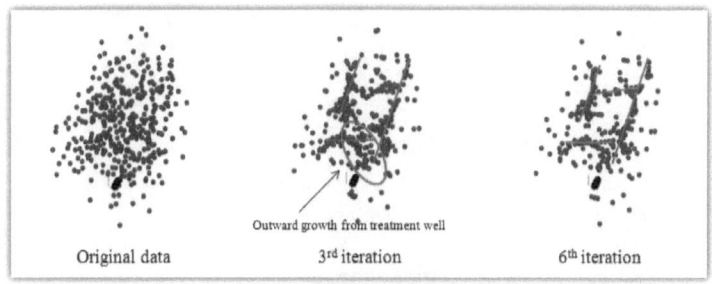

Original data   Outward growth from treatment well   3rd iteration   6th iteration

Fig. 4. Statistics Graphics related with carjackings.

To implement the application is installed in operating system devices with Android 2.2 or higher, which tests the system in different areas of the city based on the previously research related with Cultural Algorithms on Urban Transport [5], by answering a questionnaire of seven questions to all users after a scholar semester have elapsed since installing the application, the questions are to raise awareness of the performance, functionality and usability of the system, the demonstrate use of this application is shown in figure 5. To understand in an adequate way the functionality of this Intelligent Tool, we proposed evaluate our hybrid approach and compare with only data mining analysis and random select activities to protect in the city, we analyze this information based on the unit named "époques", which is a variable time to determine if exist a change in the proposed solution according at different situation of carjackings attacks.

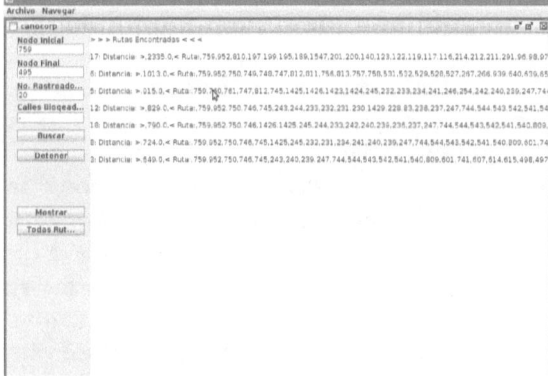

Fig. 5. Hybrid Intelligent Application based on Cultural Algorithms and Data Mining.

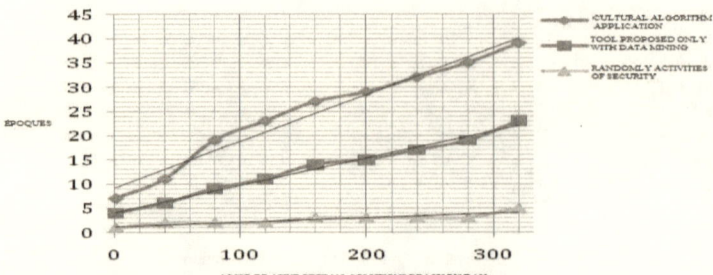

Fig. 6. Solutions proposed to carjacking problem: (blue) hybrid approach; (red) only using data mining analysis and (green) using randomly actions to improve the safety of the users.

We consider different scenarios to analyze during different time, as is possible see in the Figure 6, and apply a questionnaire to a sample of users to decide search a safety place in the City, when the users receive information of carjacking (Data Mining Analysis) try to improve their space of solution but when we send solutions amalgam Cultural Algorithms and data Mining was possible determine solutions of security by the users and describe the real situation of possible danger, the use of our proposal solution improve in 82% against a randomly action and 28% against only use Data Mining analysis the possibilities of recommend leave a place with possible carjacking attacks, these messages permits in the future decrease the possibility of suffer a fatal carjacking.

Conclusions

With the use of this innovative application combine Cultural Algorithms and Data Mining based on a mobile dispositive is possible determine the places where is possible occurs a carjacking in Juarez City by an alert sent to a mobile device with GPS, providing statistical information through a Web server that returns the level of insecurity in the area consulted [4]. The future research will be to improve the visual representation of carjacking to a social networking to this we proposed an Intelligent Dyoram with real on time information of each one of their integrants. The most important contribution is prevent more deaths in the city because drive in a unsafe place on a wrong time and suffer an carjacking attack, our future research is adequate the information to actualize from the central server of security of the city, to the users, considering that 12750 people died during this last six years in Mexico related with this cause of death, this innovative application is possible to use in another great cities in Latin America, Africa, Oceanía and Asia with similar problems of insecurity, this Intelligent Tool will be used by different kind of people whom suffers 185 cases during 2012. In addition this application will be used as Recommender System when travel to another societies as Barbados, Basque Country, Denver, Equatorial

Guinea, Hawaii, Irán or inclusively North Korea [7] and explain different scenarios according time and location.

References

1. Alonso F. Ignacio. "Las coordenadas geográficas y la proyección UTM", Universidad de Valladolid, Ingeniería cartográfica, geodésica y fotogrametría. Febrero 2001.

2. Andreu R. Alejandro. "Estudio del desarrollo de aplicaciones RA para Android," Trabajo de fin de Carrera. Catalunya, España, 2011.

3. Barrera J. Orlando, "Sistema de Información Geográfica Móvil Basado en Comunicaciones Inalámbricas y Visualización de Mapas en Internet," M.C. tesis, Ensenada, Baja California, México, 2011.

4. Cáceres, Amalia. "Sistemas de Información Geográfica" Profesorado en Geografía. Instituto Formación Docente P.A.G., 2007.

5. Cruz, Laura; Ochoa, Alberto et al.: A Cultural Algorithm for the Urban Public Transportation. HAIS 2010: 135-142.

6. Glass, Steve; Muthukkumarasamy Vallipuram; Portmann, Marius: The Insecurity of Time-of-Arrival Distance-Ranging in IEEE 802.11 Wireless Networks. ICDS Workshops 2010: 227-233.

7. Souffriau, Wouter; Maervoet, Joris; Vansteenwegen, Pieter; Vanden Berghe, Greet; Van Oudheusden, Dirk: A Mobile Tourist Decision Support System for Small Footprint Devices. IWANN (1) 2009: 1248-1255

Determining Anthropometry related with Fencing using Social Data Mining

Alberto Ochoa[1], Arturo Hernández[2] & Halina Iztebegovic[3]

[1]   Instituto   de   Ingeniería   y   Tecnología   (Departamento   de
Computación); UACJ, México.

[2] CIMAT, Guanajuato; México.
Technical Montenegro University, Montenegro.
E-mail: cbr_lad7@yahoo.com.mx

Abstract—To assess the differences in anthropometric parameters, body fat, body mass index (BMI) and body density induced by sport—specific morphological optimization (adaptation) between two samples (Juarez City University and Université Quebecoise Au Montreal) of male fencing players. The survey included a total of 160 male fencing players, all members of Universitary teams of Fencing. The sample from Juarez City consisted of 95 players (71.9% of target population) aged between 18 and 30 years, and the sample from Montreal included 65 players (50% of target population) aged between 19 and 29 years. Trained and qualified anthropometrists performed the measurement under standardized experimental conditions and in accordance with the procedures described by the International Biological Program. They measured 23 anthropometric variables reflecting basic human body characteristic described by skeletal bone lengths (total leg length, total arm length, hand length, foot length, and height), breadths (hand at proximal phalanges, foot in metatarsal area, biacromial, biliocristal, biepycondylar femur, biepycondyar humerus, and radio—ulnar wrist breadth), girths (chest, arm, forearm, thigh, and calf girth), skinfold thickness as a measure of subcutaneous adiposity (triceps, subscapular, axillary, calf, and abdominal skinfold thickness), and mass. Additionally, estimates of body mass index (BMU), body density, and percentage of body fat were calculated from the primary measures.

Index Terms—Anthropometrics, Social Data Mining.

I. INTRODUCTION

Fencing is a tactical sport, a sport game with an aerobic-anaerobic character. Variations in body size due to environmental influences are much larger than those resulting from genetic differences [1]. A trend of increasing body size and faster growth rate has been noted in industrialized countries since the middle of the 19th century, especially in the first half of 20th century [2, 3]. This positive secular trend has largely been attributed to improving living conditions, nutrition, and control of infections [4-6]. The secular trend of increased stature observed during the last century amounted to 1.3 cm per decade by the end of childhood, 1.9 cm in mid-adolescence, and 0.6 cm at young adult age [7]. Different effects of sport activities (sport training) on growth and development have been summarized different publications and textbooks [8-10]. The athlete's anthropometric dimensions, reflecting body shape, proportionally, and composition [11, 12], plays a significant role in determining the potential for success in sport [13]. Distinctive anthropometric characteristics come about by natural selection of successful athletes over successive generations and/or by an adaptation on the training demands within the present generation. The "final" body shape and composition in a given sport results from a phenomenon called "sport morphological optimization" [14]. A number of differences in players' body morphology and composition [15, 16] due the environmental changes in general, and changes within the game of Fencing itself [17, 18] could be expected. The aim of this study was to determine the size and direction of changes in anthropometric characteristics of two different Fencing players in two societies.

II. Subjects and Methods

The sample consisted of 160 male Fencing players from two samples. Anthropometric measures of 95 players (71.9% of the target population) were taken in Juarez City (Tritons team), and of 65 players in Montreal. The age range of the first sample at the time of measurement was 18-32 years (mean ± standard deviation, SD, 21.1± 4.0 years), and the age range of the second sample (50% of the target population) was 19-29 years (mean ± SD, 21.8 ± 3.8 years). All participants were clinically health without morphological aberrations. The only inclusion criterion was participation in at least one University team in the year of measurement. There was no overlap between the two groups.

Trained and qualified investigators performed all the measurements, using standardized procedures recommended by the International Biological Program [19]. A medical balance was used with a precision of 0.1 kg, Martin anthrop meter with a precision of 0.1 mm, small sliding caliper with accuracy of 1 mm^2, and synthetic length measuring tape with accuracy of 1 mm. Anthropometric status of subjects was determined on the basis of 23 anthropometric measures. Katch and McArdle formulae were used to estimate the body fat percentage on the basis of measurements performed [20].

Data for each sample were presented as mean ± standard deviation (SD). Analysis of variance was used to test the differences between the two generations. Differences were also calculated as z-scores (z^2). Measures of the two samples were rescaled according to the formula: $z_2 = (r_2 - x_1) / sd_1$, where r_2 was the result of Montreal and x_1 and sd_1 were mean ± SD for Juarez people. SPSS statistical software, Ver. 11.0 was used for all statistical analyses.

The two samples significantly differed in almost all anthropometric measures and indices (Table 1, Fig. 1) except chest girth, arm girth, mass, foot length, and foot breadth. Comparison of length measures between the samples, ie, the leg length, arm length, and height, showed statistically significant positive trends. N exception was the hand length, where a significant decrease in the mean value was observed. Breadth measures, including biiliocristal breadth, biepcondylar femur, biepcondilar humerus, and wrist breadth, significantly decreased. Only the biacromial breadth increased in the second sample. The measures of soft tissues, such as girths and skinfolds, generally decreased. There were significant decreases in forearm girth, thigh girth, and claf girth. In skinfolds, estimators of the amount of subcutaneous adipose tissue measures of triceps, subscapular axillar and abdominal skinfgolds showed significant decreases, and only the calf skinfold was increased. Body mass index (BMI) and estimated body fat percentage decreased significantly, whereas body density increased. While the mean height significantly increased, body mass remained statistically unchanged.

III. Discussions

Our study showed that body morphology and composition of Fencing players significantly is different in Juarez City and Montreal. The players has longer limbs, and smaller breadth and girth of most anthropometric parameters measured. Their bodies were taller but more slender with wider shoulders and thinner waist. It seems that a change in the body composition of Fencing players has been accompanied by changes in the body shape. Body adiposity estimate, based on the skin folds, was significantly lower. In the second sample, estimated percentage of body fat and BMI was lower, and body density higher. Since there was no difference in the body mass between the two samples, it seems that the same body mass was achieved by the increase in the muscle and bone mass on account of less dense body fat, which decreased, all this characteristics has influence on the genetics of the societies.

Table 1. Analysis of variance of differences in 26 anthropometric measures, body mass index (BMI), body density, and body fat between Male Fencing players of two different samples.

Parameter (mm)	Anthropometric measures (mean ± SD, range) of water polo players in the year		d*	F (ratio)	p
	Jurnis Srty (n = 95)	Nordeegrs (n = 65)			
Leg length	1,056.3 ± 41.8 (967-1,168)	1,073.3 ± 38.4 (986-1,177)	-16.98	6.801	0.010
Total arm length	802.9 ± 30.9 (704-884)	831.5 ± 34.5 (748-907)	-28.59	30.077	0.001
Hand length	200.3 ± 8.9 (172-227)	186.2 ± 8.1 (167-204)	14.14	104.847	0.001
Foot length	280.5 ± 11.0 (258-312)	280.3 ± 10.6 (251-296)	0.24	0.19	0.890
Height	1,858.6 ± 52.7 (1,741-2,000)	1,895.9 ± 50.2 (1,789-2,018)|	-37.31	0.131	0.001
Hand breadth (proximal phalanges)	81.9 ± 5.6 (69-96)	84.1 ± 4.1 (76-93)	-2.16	7.066	0.009
Foot breadth (metatarsalis)	102.8 ± 6.2 (87-118)	101.3 ± 5.3 (91-115)	1.53	2.681	0.104
Biacromial bredth	420.6 ± 19.5 (372-468)	437.3 ± 13.3 (401-462)	-16.70	36.174	0.001
Biiliocristal bredth	297.1 ± 14.9 (265-353)	285.2 ± 15.8 (256-330)	11.91	23.408	0.001
Biepycondilar femur	99.3 ± 5.2 (90-115)	96.5 ± 4.5 (88-108)	2.80	12.428	0.001
Biepycondilar humerus	73.1 ± 3.4 (65-80)	65.7 ± 5.8 (54-79)	7.38	101.413	0.001
Wrist breadth (radio-ulnar)	60.6 ± 2.8 (53-68)	58.1 ± 2.6 (51-63)	2.48	31.925	0.001
Chest girth	1,030.5 ± 45.5 (932-1,154)	1,039.4 ± 51.1 (942-1,156)	-8.88	1.330	0.251
Arm girth (relaxed)	328.3 ± 20.8 (282-385)	324.9 ± 17.3 (293-381)	3.34	1.138	0.288
Forearm girth	282.1 ± 11.4 (256-312)	273.4 ± 12.3 (241-300)	8.67	20.861	0.001
Thigh girth	601.3 ± 28.3 (533-682)	565.0 ± 26.2 (507-631)	36.29	67.589	0.001
Calf girth	389.1 ± 15.9 (354-431)	375.7 ± 14.2 (341-413)	13.43	30.018	0.001
Triceps skinfold	9.3 ± 2.8 (4.8-19.3)	8.2 ± 2.7 (4.35-16.52)	1.10	6.219	0.014
Subscapular skinfold	11.0 ± 3.2 (7.2-22.8)	9.0 ± 2.3 (6.28-16.70)	2.02	19.426	0.001
Axilar skinfold	9.1 ± 3.7 (4.3-20.3)	7.3 ± 2.8 (4.38-20.87)	1.78	11.030	0.001
Calf skinfold	8.0 ± 2.2 (3.8-14.8)	10.6 ± 3.1 (5.78-20.20)	-2.60	38.834	0.001
Abdominal skinfold	13.4 ± 5.6 (5.5-27.5)	10.6 ± 4.5 (5.22-29.60)	2.87	12.026	0.001
Mass (kg)	85.2 ± 7.3 (65.6-107.2)	85.9 ± 6.9 (73.0-104.0)	-0.74	0.419	0.518
Body mass index (BMI)	24.7 ± 1.7 (20.0-30.1)	23.9 ± 1.4 (21.31-28.91)	0.76	8.798	0.003
Body density (against water)	1.07 ± 0.01 (1.06-1.1)	1.08 ± 0.01 (1.06-1.08)	0.0038	13.781	0.001
Body fat (%)	11.1 ± 3.0 (6.9-19.2)	9.4 ± 2.4 (6.4-17.5)	-1.65	13.693	0.001

IV. Evaluation of Results

These differences maybe explained by the changes in Fencing and changes in environmental conditions. The playing rules of the game over the World were subject to several changes. The variants of sport with less time to escaramuza. Accordingly, there are less physical contacts between opponent players during the plactice the ratio of vertical to horizontal posture to attack changed in favor to the horizontal one, and there are more contact points [18]. Consequently, the volume and intensity of the training also considerably increased.

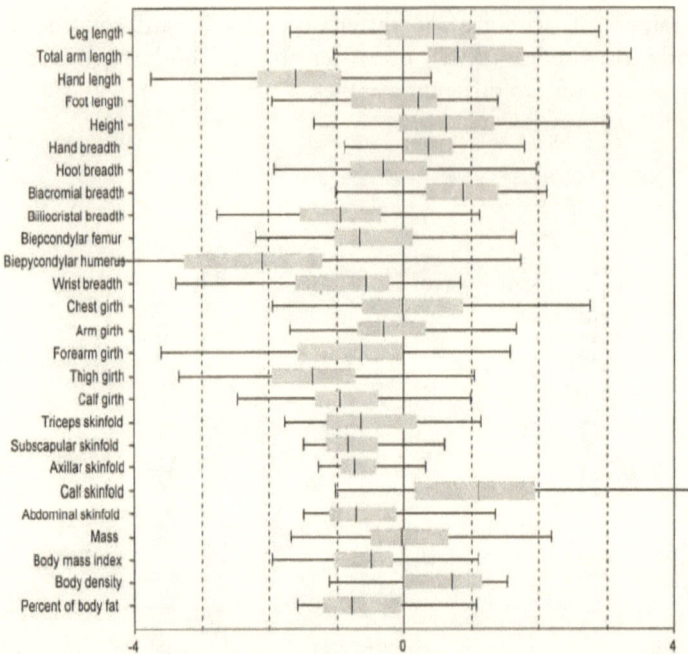

Figure 1. Differences in 26 anthropometric variables, body mass index (BMI), body density, and percent of body fat between two samples of male Fencing players. The cross-sectional study was based on two measurement point. Box-plot of measures at the 2nd measurement point is rescaled as z-scores to the 1st sample. Full line in the box –arithmetic mean, box- 2 standard deviation, whiskers –minimum and maximum. Zero on the x-axis denotes the mean of all 26 anthropometric variables, with up to 4 standard deviations (-4 to 4).

The significantly shorter hand length in Juarez City in comparison with Montreal was most probably the consequence of the changes in playing conditions and rules of the game. Changes in rules of the game which is more competitive promoted the use of technologically improved espadas, sables & floretes allowing it to retain its stable characteristic for the full duration of the game. Conversely, in Juarez City played the game with different accessories; consequently, the characteristics changed during the course of the game as it became heavier. It seems that shorter hand in the Fencing players measured in Juarez City was a disadvantage for better manipulation and control of the arms.

The secular changes, ie, a more rapid growth and development, higher mean stature, and body mass have been noticeable in Developed countries and elsewhere for more than a century. Coefficients of increase in the stature per decade (cm/decade) differ among countries, from 0.3 in

Norway and Sweden to 1.9 in Slovenia (6). The average secular trend coefficient for Europe is 1.2 cm/decade.

Thus, the lack of differences in body mass and a greater than expected increment in height imply some other sources of variation besides the already established population secular trend. Due to similar training histories and psychological attributes, traits other than anthropometric characteristics contributing to success, such as skill level and physical fitness, will tend to optimize similarities among athletes. Given the possibilities of influencing body shape and body composition, it is obvious that anthropometric characteristics are of paramount importance for the selection and success of new players. The observed trend can be only partially explained by a population secular trend. It is at least a twofold dynamic problem. The question on the one side is what makes a successful athlete, and on the other, how the training process and selection pressure, taken together, transform or change body characteristics. Besides a secular trend, which is obviously present in any given population/nation [6, 7] athletes are additionally influenced by the training and selection dynamics [9, 10]. The characteristic body shape and composition of successful elite athletes, in the long run, are the result of the selection process of ever changing competitive conditions in sport. Fencing players have been under pressures from intensive training and selection procedures over a number of years. Hence, the sport morphological optimization /adaptation [14] must be taken into account. Regarding the types of adaptation that have occurred, it is evident from the results in our study that the body mass of Fencing players is in the category of "absolute sports morphological optimization" , whereas their height is in "open upper end optimization" [14].

Figure 2. A Brazilian Fencing athlete with the adequate anthropometry.

The limitation of our study was that the anthropometric measurements were performed only twice. Therefore, possible extrapolations or anticipations of trends have to be made with utmost caution. Secular trends in anthropometry are the result of cross-sectional studies of populations. Furthermore, secular trend analysis and identification in particular population of athletes is additionally limited by the fact that the same athletes are usually the constituents of two societies.

Comparison between anthropometric measures of the two samples of Fencing players revealed a positive trend in body skeletal measures and negative trend in body adiposity measures. Most noteworthy differences (d) were an increase in height (d $=37.3$ mm, p$\leq$0.001), decrease in estimated body fat (d$=-1.65\%$, p$\leq$0.001) accompanied by higher body density (d$=0.001$, p$\leq$0.001), with no significant difference in body mass (d$=-2.74$ kg, p$=0.518$).

V. Conclusions

Our study may, nevertheless, provide a good insight into the anthropometric changes in Fencing players. We used multiple anthropometric variables, rarely covered to such an extent in other similar studies. The sample because of range of anthropometric measures, this study is unique and may be a challenge for further investigation of the population secular trend and sport morphological optimization phenomena, with special consideration given to their possible interaction. We believe that successful prediction of future athletes' body shape and form should be sought also in the domain of complex systems theory [21].

Anthropometric characteristics of Fencing players have changed over the analyzed of two samples. Body shape changed in terms of greater height and more elongated limbs, with thinner waist and broader shoulders. Body mass remained is very different. Muscle-to-fat mass ratio increased in the first sample. The observed changes may be a consequence of population secular trend and sport morphological adaptation (optimization).

REFERENCES

[1] Johnston FE. Environmenatl constraints on growth extent and significance. In: Hauspie R, Lindgren G, Falkner F, editors. Essays on auxology presented to James Mourrilyan Tanner. London: Castlemead Publications; 1995. p. 402-13.

[2] Tanner JM. The secular trend towards earlier physical maturation. Belg Tijdschr Geneesk. 1966; 44: 524-39.

[3] Ljung BO, Bergsten Brucefors A, Lindgren G. The secular trend in physical growth in Sweden. Ann Hum Biol. 1974;1:245 256.

[4] Van Wieringen JC. Secular growth changes. In: Falkner F, Tanner JM, editors. Human growth: a comprehensive treatise. 2nd ed. Vol. 3. Postnatal growth. New York (NY): Plenum Press; 1978. p. 307-15.

[5] Tanner JM. Growth as mirror of the condition of society; secular trends and class distinction. In: Demirjian A, editor. Human growth: a multidisciplinary review. London: Traylor & Francis; 1986. p. 3-34.

[6] Hauspie RC, Vercauteren M, Susanne C. Secular changes in growth and maturation: an update. Acta Pediatric Suppl. 1997; 423: 20-7.

[7] Meredith HV. Finding from Asia, Australia, Europe, and North America on secular change in mean height of children, youths, and young adults. Am J Phys Anthropol. 1976; 44: 315-25.

[8] Malina RM. Research on secular trends in auxology. Anthropo Anz. 1991; 48: 209-27.

[9] Malina RM, Bouchard C. Growth, maturation, and physical activity. Champaign (IL): Human Kinetics Book; 1991.

[10] Borms J, Hebbelinck M. Review of studies on Olympic athletes. In: Carter JE, editor. Physical structure of Olympic athletes. Part II: kinanthropometry of Olympic athletes. Medicine and Sport Science Series, XVIII. Basel: S. Karger; 1984. p. 7-27.

[11] Carter JE. The somatotypes of athletes – a review. Hum Biol. 1970; 42: 535-69.

[12] Carter JE, editor. Physical structure of Olympic atheletes. Part II. Kinanthropometry of Olympic athletes. Medicine and Sport Science, Vol. 18. Basel: S. Karger; 1984.

[15] Battinelli T. Physisque and fitness: the influence of body build on physical performance. New York (NY): Human Sciences Press; 1990.

[14] Norton K, Olds T. Morphological evolution of athletes over the 20th century: causes and consequences. Sports Med. 2001; 31: 763-83.

[15] Lozovina V. The characteristic of Fencing players in the morphological space. In: Reilly T, Watkins J, Borms J, editors. Kinanthropometry III. London: E & FN Spon; 1986. p. 215-20.

[16] Vujovič D, Lozovina V, Pavičić L. Some differences in ahtropometric measurements between elite atheletes in Fencing and rowing. In: Reilly T, Watkins J, Borms J, editors. Kinanthropometry III. London: E & FN Spon; 1986. p. 27-33.

[17] Pavičić L. Some possibilities for formal definition of Fencing game. In: Perl J, editor. Sport und informatik II. Köln: Bundesinstitut fur Sportwissenshaft GmbH; 1991. p. 37-45.

[18] Lozovina V, Pavičić L, Lozovina M. Analysis of Indicators of load Turing the game in activity of the second line attacker in Fencing. Coll Antropol. 2003; 27: 343-51.

[19] Weiner JS, Lourie JA. Human biology: a guide to field methods. IBP handbook No. 9. Oxford: Blackwell Scientific Publications; 1969.

[20] Katch FI, McArdle WD. Prediction of body density from simple anthropometric measurements in college-age men and women. Hum Biol. 1973; 45:445-54.

[21] Forrester JW. Principles of systems. Norwalk. Productivity Press; 1968.

Decision Support System to reduce dropout rates in Chihuahua's minorities

Raquel Santamaría[1], Alberto Ochoa[1], Flor Urbina[1] & Sandra Bustillos[1]

Juarez City University[1]

This research explains the importance of adequate different aspects related with Strategic Planning. We focus our analysis on a specific problem related with reduce dropout rates based on decision support systems under uncertainty. To this end, we performed surveys to gathering information about salient this problem using Data Mining techniques to profile a number of behavioral patterns and choices that describe social behaviors. We will define the terms "Data Mining" and "Decision Support System" as well as their contrast and roles in modern societies. Then we will describe innovative models that capture salient variables of modernization, and how these variables give raise to intervening aspects that end up shaping behavioral patterns in social aspects. We will describe the data mining methodologies we used to extract these variables related with this problem to each minority, including the analysis of diverse surveys, and provide a comparative analysis of the results in light of the proposed innovative public policies. On the rest proposed chapter, we will describe how the use of these techniques can be extended to provide a means for identifying potential social public politics. More particularly, we make allusion to behavioral pattern recognition mechanisms that would identify the importance of use Data Mining. We will close with concluding remarks and extended discussions of our approach and will provide general guidelines for future work in diverse application domains, including further analysis on how those public politics organize and operate. Our literature review will include cases of implementation of correct public politics, and some issues, challenges, opportunities, and trends about this educational problem.

The proposal of this chapter is to explain the importance of use Data Mining in a wide variety of activities related with this problem, which will be online and involve social networkings. This technique will be useful for answering diverse queries after gathering general information about this given topic. This kind of behaviors will be characterized by take a real implementation of a correct solution, each one of these taking diverse models or multi agents systems for adequate this behavior to obtain information to take decisions that try to improve aspects very important of their scholar lives.

1. - Description Dropout, and its effect on Minorities of Chihuahua.

Chihuahua, the largest state of the Mexican Federation and located in the northern province, is a multicultural society like the other states near the borders, from the mid 80's, waves of time coming from Zacatecas, Durango, Coahuila & Veracruz, with ethnic minorities represented by the group are known as the Raramuri and two different religious groups (Mennonites and Mormons) result in a diverse mosaic of living should not always in the best conditions to develop in the educational environment An important aspect to note is that the dropout exists in each of the three grade levels (kindergarten, primary and secondary), but affects a higher proportion of children of the four minorities, this research seeks to explain the reasons that justify the accumulation of factors that determine these minority groups are more vulnerable to dropping out.

Dropout is defined, the partial or total interruption of studies in the first three educational levels caused by a situation that prevents determining the parents or guardians fail to send the boy or girl to school, because in these three levels of education, compulsory, secular, and is considered an offense of a serious nature not to send a child of this age between 3 and 15 years to receive basic training. A new generation of children and young people born in Chihuahua but Federation Immigrant parents must carry the stigma of being different in their own company, along with three other minorities face every day related to the lack of a public policy to safeguard the ethnic, cultural, social and religious must conform in every classroom across the state of Chihuahua.

To determine in a more objective, this comparison, we decided to use Social Data Mining, especially Sociolinguistic analysis to evaluate the comments made in a series of group interviews conducted in the State of Chihuahua in January 2012 and the other part of an investigation of educational indicators for the four minority groups, most of these indicators are described in an informal way, the details of dropout by children and young people of ethnic minorities and even many times you can access data of age, residence, socioeconomic status and social class.

2. - Methodology

The name of Data Mining, this related to the similarities between searching for valuable information in large databases - for example, find information on trends in social behavior in large amounts of stored Gigabytes - and the mining of a mountain to find a vein of valuable metals. Data mining automates the process of finding predictive information in large databases (see Figure 1). Questions that traditionally require intensive manual analysis can now be directly and easily answered from the information [1].

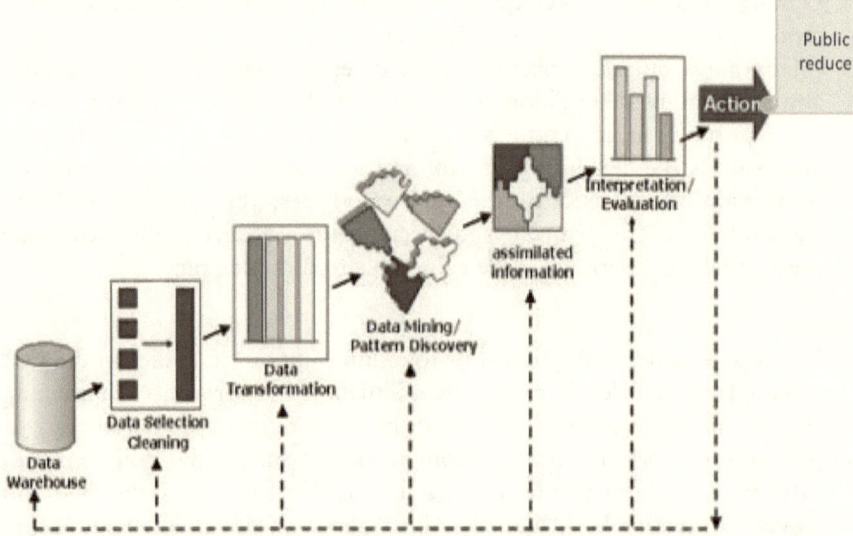

Figure 1. Data Mining Process. The information society is within Database, which are cleaned and stored in a Data Ware House, they are undermined by a cycle backward through the selection process and evaluation of patterns respectively.

The first step in developing an appropriate projection of numerical prediction based on Data Mining gives was to determine the proportionality between the different groups by category for the State of Chihuahua and determine the amount of school-age individuals (primary and secondary) which represent 28% of the population (see Figure 2).

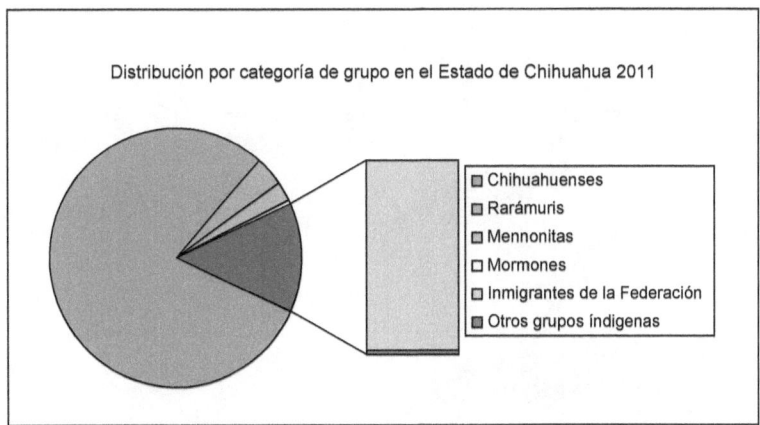

Figure 2. - Distribution by category of social groups in Chihuahua in 2011.

One of the most important aspects that can be observed in Chihuahua is the diversity in cultural patterns established for each category, a perspective view of everyday life from a different manufacturer and amalgamated for each. To determine the number of school children for the first three school levels, we use a population pyramid for the city's most populous state, in this case is Ciudad Juarez with an estimated population of 1,017.364 inhabitants (estimate for 2012) and determine the range of children and youth related to an age range of 3-15 years as shown in Figure 3, in that society are children and young people who have arrived from other societies of the Federation, who are more vulnerable to leave school in the first three educational levels in the last school year 2010-2011 through 87500 students in these three school levels, and not re-enroll for next school year.

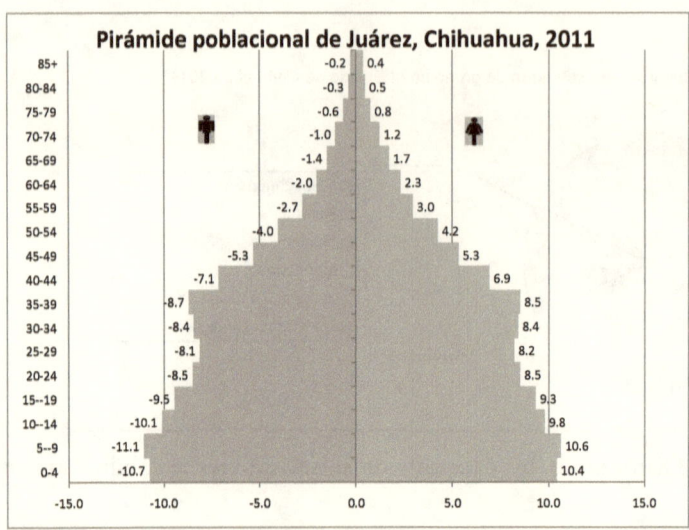

Pirámide poblacional de Juárez, Chihuahua, 2011

Age	Women	Men
85+	-0.2	0.4
80-84	-0.3	0.5
75-79	-0.6	0.8
70-74	-1.0	1.2
65-69	-1.4	1.7
60-64	-2.0	2.3
55-59	-2.7	3.0
50-54	-4.0	4.2
45-49	-5.3	5.3
40-44	-7.1	6.9
35-39	-8.7	8.5
30-34	-8.4	8.4
25-29	-8.1	8.2
20-24	-8.5	8.5
15--19	-9.5	9.3
10--14	-10.1	9.8
5--9	-11.1	10.6
0-4	-10.7	10.4

Figure 3. - Population Pyramid in Ciudad Juarez with the distribution of women and men with their respective age groups.

Chihuahua have coverage rates of 63.9% Preschool, Primary 99.9% and finally 75.9% to High School, which puts us at # 29 within the Federation, this because the lack of coverage in rural areas of the state , discrimination and social isolation in the group of immigrants from the Federation, including children and young second generation, child labor in small communities of less than 1500 inhabitants, malnutrition and poverty indicators in indigenous groups, and cultural aspects associated with religion, including a collective imagination that determines a child study only one year of preschool (Chihuahua placing at number #26 in this category), this is detailed in Figure 4, which shows the rate of enrollment in advance regular and its comparison by state and age.

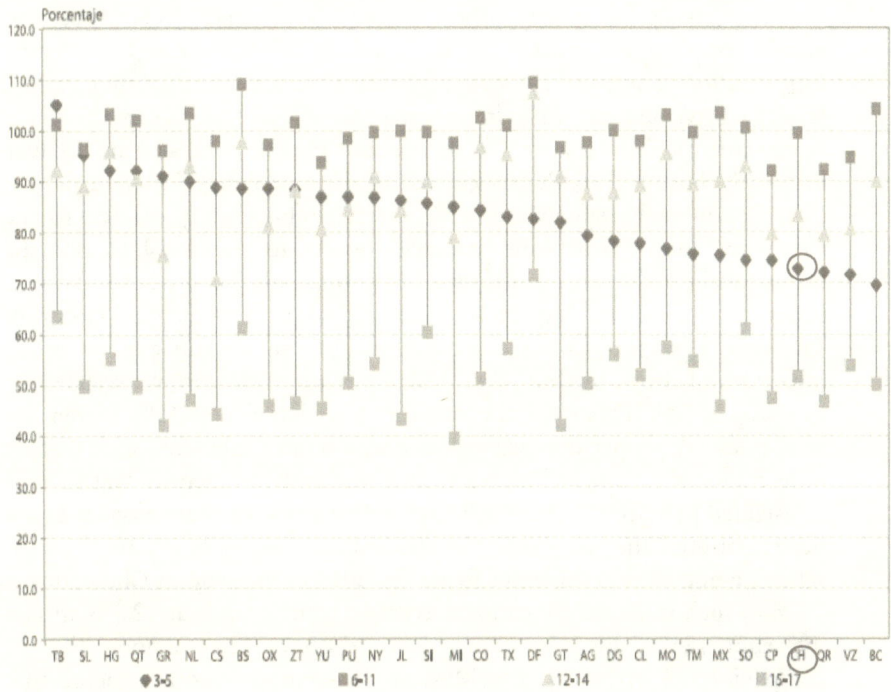

Figure 4. - Registration fee with regular progress by state and age group (2007/2008)

To understand, in a more detailed existing problems of school dropouts in Chihuahua, we use a comparison between the countries of the Organization of Economic Cooperation Development (OECD), we decided to determine whether a relationship exists between the size of their respective minorities, school dropouts, to somehow be able to justify if there is any relationship between these two social attributes to justify what happens in the State of Chihuahua. Carefully analyzing OCDE Databases, we realize that there is no plausible relationship between the minority population and dropout, in fact for some companies these minority groups have to seek federal support representative thereof, as with the inhabitants Rapa Nui in Chile, the Sino-Brazilian in Brazil, Australia and Keeling Muslim Community in New Zealand Rarotongués, these groups have a constitutional protection to access government support in the first three levels of education and not abandon them. Regarding the comparison with Chihuahua and strictly followed in an analysis of Social Data Mining, characterized with the OECD data, the index Welfare of Life for each of the 34 societies that comprise the Agency (see Figure 5), analyze the comparative comprised of 11 attributes:

Housing (Access to quality room) Income (economic), Jobs (Access to gainful employment based on the studies), Community (Community Organization Society), Education (Access to education system at all levels), Environment (Environmental Education and care for the environment), Governance (A democratic government and its citizens aware of), Health (Health services for all the Company), Life Satisfaction (Satisfaction with expectations life), Safety (Public Safety offered to the Company) and Work-Life Balance (work-life relationship on the economically active population) we found that only Turkey has indicators lower than those of Mexico and Chihuahua as a Society is closer to Turkey with respect to these social attributes, so we can expect similar indicators of dropouts, but when making that comparison, we realize that the drop is 3% higher for the majority population and 11% higher for minorities. A proper comparison of dropouts in Chihuahua with respect to its minority groups is possible to analyze, if we can identify that there is a sustained pattern of behavior from the first grade level analyzed as is the preschool until the last student in the shape lower, as might be the high school, grade level in the latter figure for 2011 in the state of Chihuahua is 19.7% which is above the national average, which stands at 12.3%, in fact the entity ranks third in high school dropout level among the 32 companies that make up the Mexican Federation. Another factor that must be analyzed, is the violence in Ciudad Juarez, which is in turn the most populous city in the state of Chihuahua, accounting for 37% of state residents since 2007 presents homicide rates above the national average and since 2008 has been placed with one of the highest rates worldwide in 2010 taking a total of 3116 deaths, and an accumulated 19,750 deaths for July 2012 (see Table 2), this has generated a mass exodus that has affected the planning horizon of the City and therefore the growth of enrollment in the first three school levels, because the people under threat and the best economic potential escape from the city, a total of 47.415 families - This would explain in part the loss of at least 450,000 inhabitants, whom have left the city as seen in Figure 6

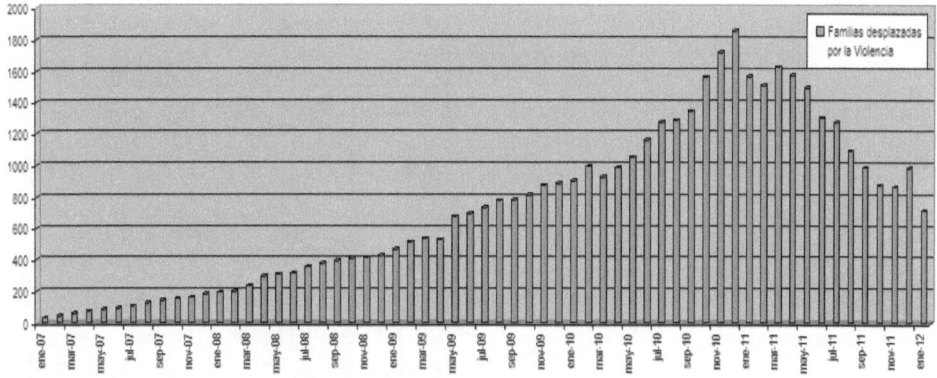

Familias desplazadas por la Violencia

Figure 5. - Families leaving the City due to insecurity. Source: Data from a CIS Intelligent Numerical Prediction Model developed by the authors.

Table 1. Deaths by year, sex ratio and average age established by gender.

Año	Muertes con violencia*	Proporción Sexo	Promedio de edad
2012	1981**	F: 21.5%; M:78.5%*	F: 21.4; M: 20.3*
2011	2047	F: 27.3%; M:72.7%	F: 23.7; M: 21.6
2010	3116	F:16.9%; M: 83.1%	F: 24.2; M: 22.3
Total	12585		

* Source hemerographic by the CIS and the Citizen Observatory UACJ both.

** Using a numerical prediction model, adjusted for safety variables and change of government at the national level, 717 deaths per August 31, 2012.

Whereas a study directly with minority students, we take into consideration that were designed with several focus groups, which are concentrated in Table 3, this collects several issues related to discrimination and social isolation among minorities analyzed, these samples formed by 387 individuals (157 women and 281 men) included children of primary level (fifth and sixth grade) and secondary (three years): Sample 1 - Mennonite: 82 (F: 36, M: 46), Sample 2 - Mormons: 62 (F: 25 F: 37), Sample 3 - Raramuri: 87 (F: 24 F: 63) and Sample 4 - Immigrants of the Federation: 167 (F: 72 F: 85), finally and to validate the information, we encourage a mixed sample (a girl and a boy in sixth grade in the Midwest (Morelos) that determine how they perceived the

117

treatment of most if they come to live with his family to Chihuahua and as addressed in their schools to children from Chihuahua.

Table 3. - Effect of school and extracurricular activities related to the interaction with the majority group.

	Muestra 1		Muestra 2		Muestra 3		Muestra 4		Muestra 5	
	F	M	F	M	F	M	F	M	Mixto	Mixto
N	36	46	25	30	24	63	72	85	2	2
Asistencia a cumpleaños de niños de la mayoría	78%	85%	75%	81%	75%	71%	66%	66%	85%	100%
Borrar el pizarrón	91%	84%	92%	91%	94%	93%	87%	86%	85%	100%
Cantar villancicos navideños	35%	29%	18%	44%	33%	59%	28%	31%	85%	100%
Compartir música y videos con sus compañeros	72%	89%	89%	91%	81%	77%	80%	74%	80%	80%
Contar historias desde su perspectiva de minoría	27%	33%	39%	65%	97%	62%	15%	31%	100%	100%
Declamar una poesía	19%	14%	11%	16%	19%	21%	10%	6%	80%	100%
Incidencia de la Educación Ambiental en sus vidas	69%	91%	71%	79%	71%	87%	64%	66%	100%	100%
Intercambio lúdico con niños de la mayoría	19%	57%	32%	40%	24%	66%	19%	20%	70%	100%
Motivación de sus maestros para continuar estudios	83%	67%	86%	74%	76%	70%	80%	60%	80%	100%
Organizar el día de muertos o Halloween	88%	93%	97%	98%	90%	91%	98%	91%	85%	100%
Organizar un Convivio	15%	21%	22%	28%	37%	78%	45%	59%	70%	100%
Organizar una fiesta del Día de la Amistad	25%	50%	43%	63%	38%	66%	24%	29%	75%	90%
Participación en Clase	61%	87%	75%	86%	71%	80%	73%	69%	80%	100%
Participar en la Escolta	4%	9%	7%	16%	29%	33%	1%	3%	70%	100%
Participar en un baile cultural	34%	31%	39%	21%	38%	37%	25%	26%	70%	75%
Participar en un intercambio de regalos	83%	93%	82%	98%	76%	93%	56%	66%	85%	100%
Participar en una Pastorela	31%	48%	29%	56%	48%	71%	21%	29%	80%	80%

	%	%	%	%	%	%	%	%		
Preparar el periódico mural	84	87	91	80	94	92	89	91	70%	90%
Recibir presentes en sus cumpleaños	80	72	71	86	71	91	76	66	90%	100%
Recibir visitas cuando están enfermos	45	47	43	52	29	49	37	34	75%	85%
Sentirse apreciados por sus compañeros	48	45	75	88	52	80	47	57	80%	100%
Ser el jefe de grupo	22	22	21	26	29	27	14	11	75%	100%
Sugerir actividades para la graduación	81	77	89	70	76	70	79	46	80%	80%
Supervisar a otros compañeros en una actividad	11	17	14	44	19	60	6%	6%	60%	95%
Suma de Reconocimiento Social (Media)	4.7	5.1	7.1	7.4	5.9	6.1	6.2	6.4	7.9	9.2

Source: State Committee, Chihuahua Human Rights 2010.

The use of data mining on social issues has proven to be a key task to verify if there are trends of perceived discrimination for the four minorities by the majority social group in a situation of violence in common, we find variations depending on usage of social blocking actions, social isolation and school violence (bullying), see Table 3, we can see how any of the four minorities get half more than 8 on a scale of 1 to 10, although children who reach Mormons to be more appreciated by most of his fellow workers, who are most have tried to be tolerant of others.

3. - Analysis of the sample related to dropout for four minorities in the State of Chihuahua.

Whereas each of the relevant issues associated with minority analyzed developed the following equation (Equation 2), which seeks to justify the reasons for dropouts in minority, this is defined as follows:

$$\text{Dropout rate} = [\text{School Violence (Low school performance)}] * \sum_{i=1}^{n} (\text{Social Aspects Miscellaneous}) \quad (2)$$

Based on this equation, we can determine, the geospatial location, ie the ratio of each child with respect to the others of each student and determine the long-term prospects to leave school, when viewing the graph (see Figure 7) we understand that the most vulnerable group is the Raramuri, who have more adverse economic conditions and therefore school performance is more limited in the case of this Mennonite group varies according to type of congregation in participating and the location being near Cuauhtemoc dwelling children in traditional schools and those who also suffer from social blockade by their peers, for his part in the collective Mormon we see a pattern of mutual support when they have to live with children of the majority, try to organize to get together and get through the day to day.

In the case of immigrants of the Federation (see Figure 7), which. is the group with greater differences and this is their place of origin, children suffer more harassment come from Oaxaca, Guerrero and Veracruz, an intermediate group of children from Coahuila, Zacatecas and Durango together to try to negotiate with the group majority and finally the children from the most recent wave of Mexico City, Queretaro and Morelos, even become daunting in their schools, they have the highest capital of the group and tend to go to Private Schools, so you can consider this completely heterogeneous group in its relations with the majority group.

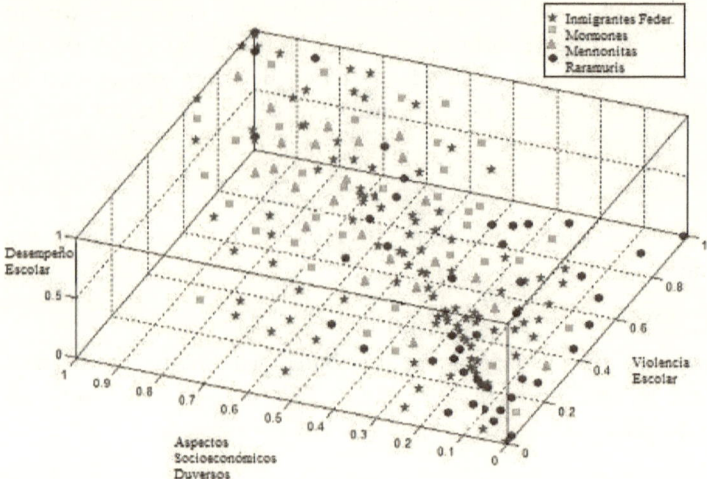

Figure 67. - Visual representation of the sample, characterizing the various socioeconomic, school violence including blocking social and school performance reflected.

Legal Basis

Seeking to regulate the situation of education, the Mexican State through the years has created and reformed laws and institutions that provide the opportunity and rights in its various areas, whether federal, state or city to study basic education and the average higher.

Corresponds First go to the supreme law of this country, the Constitution of the United Mexican States, which in its first chapter called Human Rights and Guarantees, Article Three: One of the most important items is the result of the Mexican Revolution (1910-1921), provides one of the most important rights for children, the right to education; Within the same article provides a number of points that must be addressed when referring to the issue of education and these are the following:

Everyone has the right to an education. The State-Federation, States, Federal District and Municipalities-provide preschool, primary, secondary and high school. Preschool, primary and secondary form the basic education and higher mean it will be mandatory.

The first part of the paper determines who should use this right to every individual, so that there is (or should not be) any kind of social, economic, political, etc.. and any exclusion by location, language, race, etc.. Every individual, means any person who is in our country.

According to Article 1 of the Constitution:

Article 1. In the United Mexican States all persons shall enjoy the rights recognized by this Constitution and international treaties of which the Mexican state is a party, as well as guarantees for their protection, whose exercise may not be restricted or suspended, except in cases and under conditions established by this Constitution.

Article 2 of the General Law of Education, and as a secondary law states:

Article 2. - Everyone has the right to education and, therefore, everyone in the country have equal access to the national education system, with only satisfy the requirements laid down by the general rules.

Non-discrimination

Article 1 The Constitution also provides:

All discrimination ... motivated by ethnic or national origin, gender, age, disability, social status, health status, religion, opinions, sexual preferences, marital status or any other that threatens the human dignity and is intended to nullify or impair the rights and freedoms of individuals. Definitely one of the most important in the educational aspect and based on non-discrimination, is this article which provides a clear and direct that there should be no discrimination, and recognizes the power enjoyed by all people to enforce the rights it establishes. Returning to Article 3 of the Constitution, it provides some of the educational objectives among which are the following: contribute to improved human relations, to strengthen appreciation and respect for cultural diversity, the dignity, integrity family, the conviction of the general interest of society, the ideals of brotherhood and equal rights for all, avoiding privileges of race, religion, groups, sexes or individuals, establishing the first point which would not there should be no discrimination. This education must be of quality, comprehensive education that is comprehensive, it is based on the principles of this nation, so within the same article states that will be democratic, considering democracy not only as a legal structure and political regime, but as a way of life based on constant economic, social and cultural life of people, also will be national in-without hostility or exclusiveness-attend to the understanding of our problems, the use of our resources, the defense of our political independence, the assurance of our economic independence and the continuity and growth of our culture.

Article 4 of the Constitution refers to children specifically and determine the following:

... For any decisions and actions of State and shall comply with the principle of best interests of children, ensuring their rights fully. Boys and girls have the right to satisfy their needs for food, health, education and recreation for its development. This principle should guide the design, implementation, monitoring and evaluation of public policies aimed at children.

Also in the Education Act states:

Article 8. - The criteria that guide the education that the State and its decentralized agencies, and impart all preschool, elementary, secondary, normal and other teacher training for basic education that individuals impart- be based on the results of scientific progress; fight against ignorance and its causes and effects, servitudes, fanaticism, prejudice, the formation of stereotypes, discrimination and violence especially that perpetrated against women and children and must implement State policies aimed at mainstreaming criteria in the three branches of government.

I. - It will be democratic, considering democracy not only as a legal structure and political system, but as a way of life based on constant economic, social and cultural life of people;

II. - Shall be national in-without hostility or exclusiveness - will serve the understanding of our problems, the use of our resources to the defense of our political independence, the assurance of our economic independence and the continuity and growth of our culture .

III. - It contributes to better human society, both by the elements that contribute to strengthening the student, along with an appreciation for the dignity and integrity of the family, the conviction of the general interest of society, As for the care you put into sustaining the ideals of brotherhood and equal rights of all men, avoiding privileges of race, religion, groups, sexes or individuals.

On the basis of this constitutional provision we can refer to all others and are intended to regulate education in Mexico, analyze and set a series of rights in various articles and looking as non-discrimination.

Equal opportunities

Article 2 of the Constitution. - The Nation has a multicultural composition, originally on its indigenous peoples are those who descend from populations that inhabited the present territory of the country at the beginning of colonization and who maintain their own social, economic, cultural and political or part thereof. This article provides recognition to indigenous peoples, who are entitled to equal opportunities in all aspects and as mentioned above, should not be subjected to any discrimination, either in education or any other. With the creation of this article recognize the cultural differences that exist in our country.

One of the commitments of the federation, states and municipalities is PROMOTE GENDER EQUALITY OF INDIGENOUS AND eliminate any discriminatory practice will be the responsibility of the 3 spheres of government, as provided in subsection B of the second article of the Constitution, establish the institutions and determine the policies needed to ensure the observance of the rights of indigenous peoples and the development of their peoples and communities, which must be designed and operated in conjunction with them.

Equity in Education

Of the General Law of Education presented the following:

Article 33. - To comply with the provisions of the preceding article, the educational authorities within their respective jurisdictions will conduct the following activities:

I. Take particular schools that, being in isolated, marginalized urban areas or indigenous communities, is considerably greater the possibility of delays or dropouts, by allocating better quality items to meet the educational problems of these localities;

II. - Develop programs to support teachers who provide services in isolated and marginalized urban areas, to encourage the roots in their communities and meet with the school calendar;

VII. - Will carry out educational campaigns aiming at raising the cultural level, social and welfare of the population, such as literacy programs and community education;

Article 38. - Basic education at three levels, will the adaptations required to meet the linguistic and cultural characteristics of each of the various indigenous groups in the country and scattered rural population and migrant groups.

The State will also conduct welfare programs, food aid, health campaigns and other measures to counter the social conditions that influence the effective equality of opportunity for access and retention in educational services.

The State of Chihuahua

In relation to the above items the State of Chihuahua in its legislation, has a coherence and consistency in relation to higher laws. The State Constitution places special emphasis in regard to the Indians:

Article 10. The education of indigenous peoples will be given special attention by the State. The law shall establish the necessary mechanisms to encourage that it be provided by these people and be bilingual when they so request.

It also features a special section, Title XII of the General Administration Chapter I Public Education, sections 143 and 144 determine that education is free and compulsory for all those state residents who are of school age also are states that it must be secular and again mentions the set with the Non-discrimination also refers to all the provisions of the constitutional articles mentioned above.

National Plan and State Development

The federal government and state government submitted work plans that include goals and objectives to accomplish during his administration called: National Plan and State Development, then I will refer to what relates to our subject.

Within the National Development Plan, establishes strategies related to indigenous education, first accepting the deficiencies in the quality of education and the results presented below. Within this document sets out the following obstacles:

- Shortage of bilingual teachers,

- The presence of students who speak different languages in the same group,

- Isolation and marginalization of the communities where they live.

The objectives and strategies are:

Strategy 10.3 - Strengthen efforts to integrate adult literacy and youth and adult education programs to reduce the backlog open education.

Exists within the NDP a section called Indigenous peoples and communities and which provides, among other things, the following:

The National Development Plan (Chihuahua) has a section called Ethnic groups within which we develop two chapters, the first set as indigenous peoples and communities and the second refers to the Mennonites.

As targets are presented as follows:

Objective 1. Provide mechanisms for consultation with indigenous peoples and communities that allow the government to ensure public policies consistent with their deepest needs and development.

¯ Provide food 14 000 scholarships to children and youth of Basic Education to 450 schools in 23 municipalities mountain, which will help reduce dropout rates, academic achievement will strengthen and improve the health conditions of students.

¯ Build and equip 200 school canteens for the benefit of students of basic education for mountain municipalities.

¯ Promote education programs that strengthen indigenous bilingual intercultural approach at the level of basic education teachers, including speakers of the language and the use of educational materials in indigenous languages.

¯ Establish a scholarship program with sufficient resources to support students on Higher Education. This program with a fund composed of contributions from the State Government, Federal Government, social clubs and corporate foundations.

Mennonites

Goal 2. To promote education and culture of the Mennonite community in a framework of respect for their culture, traditions and customs. Closer ties with the entire population without discrimination for integration.

2.1. Provide support to improve education while respecting and preserving the culture, while Spanish courses are implemented, computer and other programs to help its incorporated and training.

¯ Manage the provision of books, supplies, paint and furniture for schools to the official system.

¯ Provide and conduct training course for teachers in each region and the state colony, according to their needs, habits, customs and traditions with the participation of heads of Cologne, Ministers and Officials in Charge of Religious Education.

¯ Promote and manage the integration of schools into the national education system to have the recognition and official recognition.

4. - Final Considerations.

Leave school for a boy or a girl, is a dramatic situation to stop at once the hopes and dreams to access a better life in future, this experience ultimately proves to be traumatic and difficult to bear, that is why when it does, the legal levels and especially try to remedy social only through in-kind support the school situation. An inclusive public policy, would help vulnerable social groups and to be able to sustain a paradigm shift in society, which should see diversity as a bond to be part of a "bulwark of symbolic capital." There have been studies that indicate that expectations of the minorities in question are not allowed access to better jobs, leading to fewer opportunities in the future, which ultimately lead to a vicious cycle in families. Instead, the majority group have a social structure tree, branched, allowing them to interact with various people in various categories in different planes of time and space simultaneously, which is why opportunities diametrically differ according to the actions required to overcome the dropout, the proportion of the majority who see fit even engage in extremely violent attacks once they have left school represents 89% of the cases, 57% partner in the minority. With respect to Chihuahua, violent deaths from 2007 to June 2012 represent 39% of all deaths during the administration of President Felipe Calderon and can only be compared in number to the deaths of the political transition in Libya or Syria Right now, this cannot be matched from the social point of view, because the children suffer from social stigma of misunderstanding and lack of recognition by the dropout rate to be made and deserves legal recognition and hence in the society.

Based on the foregoing, we conclude that the state provides all the protection and legal support through their various laws, to gain access to free education and quality, considered as an inherent right to individual and universally, without generating distinction of sex, race, economic status, in anothers. This legislation should reflect another reality, an education to all children and adults can access without any problem. But otherwise, it is not easy to study in this country; there is a problem in implementing these laws and to exercise their rights under the Mexican

Constitution and implementing legislation and government plans. With regard to minorities there is also a protection from the State, defining the rights and laws safeguarding of discrimination or lack of equity and opportunity. In the development plans, there is a special section in this regard and establish strategies as part of public policy in this case, focusing on education. Furthermore, this study determined a section called " " citizens, which sets out the obligations of parents and society engagement for each child to attend school and be prepared academically.

References

Adderley, R., & Musgrove, P.B. (2001). Data mining case study: Modeling the Behavior of Offenders who commit sexual Serious Assaults. In Proceedings of KDD '01, San Francisco, CA.

Balkin, J. (2006). Law and liberty in virtual worlds. In Balkin, J & Noveck, B (eds.) State of Play: Law, Games and virtual worlds. New York: New York University Press.

Chen, H., Chung, W., Xu, JJ, Qin, GWY, & Chau, M. (2004). Crime data mining: a general framework and Some examples. Computer, 17 (4), 50-56.

Garcia-Ruiz, MA, Vargas Martin, M., Ibrahim, A., Edwards, A., & Aquino-Santos, R. (2009). Combating Child Exploitation in Second Life. 2009 IEEE Toronto International Conference - Science and Technology for Humanity (TIC-STH).

Curbs, R.W. (2005). Social and ethical Considerations in virtual worlds. The Electronic Library, 23, 539-546.

Schrobsdorff, S. (2007). Predator's playground. Newsweek. Available June 14, 2010, from: http://www.newsweek.com/2006/01/26/predator-s-playground.html

Identification of popularity to labeling a map-based on social networking using an Imaginary Collective and Cultural Algorithms

Daniel Azpeitia[1], Alberto Ochoa-Zezzatti[1], Rubén Jaramillo[2] & Sandra Bustilos[1]

[1]Juarez City University, México

[2]LAPEM, México

alberto.ochoa@uacj.mx

Abstract. In the present study, we describe some promising initial results from induced ontology vocabularies of labels-which a map associated with a Social Networking online were selected using Cultural Algorithms, using a model based on assumptions. Describes the utility of ontology of aspects as a supplement to a system that labeled and marked with the model and show results. Proposing a revised model using probabilistic ontologies seed to induce ontology of aspects, and describe how the model can be integrated into the logistics of labeling communities.

Keywords: Cultural Algorithms, Pattern Recognition and Decision Support System.

I.   Introduction

In recent years has seen rapid growth in the Use of labeling applications, both in the number of labeling applications associated with images or maps and in the number of users participating in communities labeling. This growth currently exceeds our understanding of how entries are made which are efficient and productive for a range of applications and users. The labeling schemes are often placed in opposition to the taxonomic models and two types are they are commonly cited:

1. User interfaces for annotation based on a closed vocabulary, hierarchical are clumsy and inflexible

2. A strict tree of concepts does not reflect its use and intention.

The first criticism is valid, but can be treated easily with dynamic ontologies better mechanisms for UI. Much of the second critical not so much an issue with the taxonomy (ontology) alone itself, but something problematic which force models users to put concepts into a single hierarchy. Much of this issue can be treated using ontologies aspects that separate the various aspects of the attempt annotation. Some of the common facet used in the annotation of the media include the location, activity or event associated with several aspects of the paint (People, plants, animals, objects, in anothers.) and especially in shared context, given the emotional response. Labels provide a simple and direct to create annotations that reflect a variety of aspects, and also provide direct means of shipping on a search. The search at least based purely labels tend to have low memory operation (This can be mitigated in part by a UI that adapts a vocabulary aligned). Moreover, when a search Initial returns a large number of results, the tags not support efficient or intuitive models of refinement of the question. In the best case, users currently can refine a search using clusters of (Statistical) concepts. Although it is sometimes useful, clustering performance that is very difficult to assess. In [7] described the distinction between polythetic clusters (in which members of the cluster share a certain proportion of system characteristics) and monothetic clusters in which all members share a characteristic in common. Because of this in [7] discussed that users can understand more easily the monothetic clusters.Furthermore, because this polythetic clusters are difficult labeling, compared with monothetic clusters that are easy to label adapting to different paradigms common interface such as guided navigation, hierarchies limited to the refinement of the question, among other. In [2] have been explored search engines based on issues, as well as in [4] and on the other hand [8] show the usefulness of the search interfaces aspects to search for images. Although some in the community marking with labels resist the taxonomy, del.icio.us even (http://del.icio.us) recently added "Packages" - while it is purely organization (the packages do not support the ontological semantics), the characteristic recognizes the challenges of scaling a model organization marking with labels. It is believed that users should not having to choose between models based purely label models with purely taxonomic vocabularies closed. We explore a model that balances technical natural language processing statistics together with domain knowledge ontology to induce can be balanced on the final answer.

The objective is a system that maintains the flexibility of the interface tagging for annotation while also benefits from the power and usefulness of an ontology aspect in the search and display interface. We present initial results of a model based on symbolic logic to the labeling system as the maps associated with Facebook. (http://www.facebook.com/profile.php?id=829140429&sk=map), demonstrating the potential for induces the technique suitable for an ontology search and display the user interface. The remainder of paper describes the approach, the test set using cultural algorithms and evaluation, coupled with a proposal for a refined model and how it fit into the logistics of a brand community by tags like occurs on Facebook

II Related Research

In [7] described a simple statistical model of logic symbolic which assumes X to Y if:

$$P(X \mid Y), \ 0.8 \text{and } P(Y \mid X) \leq 1$$
(1)

To do this in [7] applies this model of co-occurrence the terms of the concept of the documents extracted recovered to a question addressed (use a search "query" is a great help to align the domain of terms). In [1] adapted the technique to assist (ie, recorded professionally) photographs of a historical collection. The resulting taxonomies are quite noisy (ie, many of the proposed pairs the assumption is incorrect), especially given that domain vocabularies are focused by questions original. The results are included in the table below as a baseline. Despite not working properly, these models generate the taxonomy that reflects the use real, and thus it adequately to labeling applications. Many other research have experimented with using techniques to induce ontology PLN statistics including [3, 5 and 8]. a few of These [3 and 5] are dependent at least in part of speech grammar, because it is that it can only be applied on natural language contexts. Other [8] tries to match concepts to existing ontologies such as WordNet; these models may be inherently less noisy but since WordNet is based on English vocabulary standard, this can be difficult to adapt these models to dynamic and sometimes idiosyncratic vocabulary that emerges in labeling applications (eg to name events).

III. Exploratory Approach

A. Passing Fancy.

We adapted the model proposed in [7] labeling system using on Facebook, adjusting statistical thresholds for reflect the ad hoc use, and adding filters to control for highly idiosyncratic vocabulary. So X potentially including if:

$$P(X \mid Y) \geq t \text{ and } P(Y \mid X) < t$$
$$D_x \geq D_{min}, D_y \geq D_{min}$$
$$U_x \geq D_{min}, U_y \geq D_{min}$$

(2)

Wherein: t is the tendency for co-occurrence, D_x is the number of documents in which the term x occurs, and may be greater than a minimum value D_{min}, and is U_x number of users that use x in at least one entry image, and may be larger than a minimum value U_{min}. It filters the input documents (ie photos) requiring a minimum of two terms of the label to that co-occurrence was defined. A series experiments, varying the parameters t, D_{min}, and U_{min}. We sought a balance that minimized the error rate and maximized the number of pairs proposed assumption. Using more stringent values for the threshold of co-occurrence (Approaching 0.9) reduces the error rate substantially but dramatically reduces the number of pairs proposed. Useful values were between 0.7 and 0.8, below the comparable were determined empirically as in [7]. The model is more sensitive to changes in U_{min}, which D_{min}. U_{min} set to any value below 5 occurred many highly idiosyncratic in terms noisy pairs of assumption, a useful range was 5 to 20. Values D_{min} varying from 5 to 40, and proved useful as through tuning the value. Both values were increased slowly while the number of documents was increased. With the fixed input below 1 million photos, vocabulary was less stable and thus the model was sensitive to the parameters. B. Tree pruning and reinforcement.

Once you calculate the statistical co-occurrence, pairs of candidate terms are selected using the specified constraints. Then a graph is built possible relationships of father-son, and filtered out the co-occurrence with the ancestors of nodes that are logically about his father. Once the statistic is calculated the co-occurrence, selected pairs of the term of candidate using the specified constraints. Subsequently, builds a graph of the possible relationships of father-child,and filter out the co-occurrence of nodes with ancestors that are logically about his father. Ie for a given ratio of the term should be strengthened. We increased the weights of each accordingly. Finally, with each leaf on the tree you choose the best path to a root, since the weights (reinforced) of the co-occurrence for prospective parents for each node, and we join paths in trees. With systems of document sufficiently large, many of the tres result are quite large - example, cities with points of interest. There was a disproportionate number wrong paths in single-instance substructures (Singleton) and double instance (doubleton) with respect to the largest sub-structures, and filter these out jointly. This is justified since the number candidate's total tree was too big for these runs (from 500 to over 3000 candidate pairs meet with a basic assumption and the criterion of filtration), and since the ultimate goal is to provide enough structure to assist in making sense navigation and guidance through the collection. A goal improve secondary search terms by subtracting the father for images with terms of the child, and in this sense is sacrificed some recoveries certainly filtering out trees singleton and doubleton. It thinks that users will fancy treesmore sensitive to the recovery accuracy, but this aspect of the model should be evaluated in user studieslarge scale.

IV. Dataset and their analysis.

We used a snapshot of the meta-data user from Facebook with 387 places located with a respectivaly photo or more (see Figure 1). To this date, Facebook had a total of 87 million images labeled to this date, and about 37 million entries in total. About 5 million of these images were scored as "not public", so were excluded from the experimental system. The tables were modified by anonymizing for all user data (including IDs photo) and all images with less than 2 terms were filtered. This resulted in a test set of about of 7000 images. The associated vocabulary was limited to 200K and generated 5000 pairs in total (no exact number is available as we filter out some while reducing Cultural Algorithms using). Using this Evolutionary Computation technical aspects determine cultural community to assess. Among the annotations Facebook, vocabulary is inconsistent with respect to limits spelling and word (eg "Los Angeles" shows often as they can be analyzed as two terms "The" and "angels", due to some non-intuitive interface input label). In addition, many terms idiosyncratic annotation. These latter vary personal events described as a phrase label ("johnandmaryswedding" - possibly indicating some confusion.

A. Evaluation of the assumption Trees that were evaluated were manually. Each pair was marked assumption proposed as correct, inverted, related, synonymous (variants including the language in common terms as flower "/" blume "/" fleur "/" bloem "among others), or noise (entirely error). The Figure 2 shows the functionally of this research. Many of the concepts the child "Los Angeles" are suburban and points of interest; several are (possibly) related and is an example of entropy result of a statistical model. In the second example, each child node is a hyponym of "Crystal" although perhaps an art historian would create as domain model "right" in the representation use within the community from Facebook. Based on our experience and that of others (eg [6]) is presumes that the images will be recorded and retrieved at more easily possible to emphasize various aspects of the keyword: place, activity, and images. The community of Facebook also another aspect that seems to accentuate might be described best as emotion or response. In this paper the results of a proportion large shared vocabulary is tied to the names of places, although it counted with the refinements of the modelto produce more balance with other aspects. For location are considered a combination of names geographical locations as well as points of interest demarcate the most to the activity. Thus we consider "Los Angeles" reasonable parent "Chinese Theater". In the sense of a pure type of relationship does not hold, however for the utility to locate an image, entirely reasonable. In the same way "Los Angeles" may be related to, but not a parent of "muni" or "Streetfair". For generic terms such as "lake" and "park" considered instances of lakes or parks that could be reasonably children. In the images, more typical of the type relations were used: "dog" but included the specific races, "food" "kimchee" and "Creamcheese" where"Restaurant" is only related. In a large photo that shares the environment such as Facebook, relationships are less useful personal question, so it considered almost all personal names such as noise any context of the pair. Table I compares theresults related models for the assumption results.

Figure 1. View a personal site with map from Facebook.

TABLE I

COMPARISON OF RESULTS

Model	Average error / another	Relations	Correct	Relation/Aspect	Equals
A	? / 19%	23%	49%	8%	
B	105 / 0.2%	15%	10%	43%	
C	1200 +	51%	21%	5%	23%

A Sanderson and Croft7.

B Clough et al1.

C Proposed in this research..

In [7] reported a high "aspect of" numbers, and attributed this at least in part due to questions vocabulary-limiting. In [1] presents an application similar to ours and so was given a line useful background. it is believed that his model would be made better if was applied to the entire whole vocabulary of data rather than a focused question. Both in [1 and 7] appears to contain an inconsistency in the statistical model (The second term should be expressed as $P(X \mid Y) < 0.8$ and not $P(X \mid Y) < 1$, although this may be a typo in the articles.

V. Proposed Model

Initial results are very promising and give rise to additional work. The proposed model produces subtrees that generally reflect different aspects but cannot categorize concepts aspects. It has planned a series of changes to the model to address this. A. Migrating to a pure probabilistic model

Currently research to express the assumption, pruning the tree construction, and classification of appearance together in a unified probabilistic model, something similar to that proposed in [5]. A probabilistic model should be more robust, and incorporate concepts such as "the number of authors using a label "as property ladder rather than a simple threshold as in the current model. Also want to add better support for replays and spelling errors, we believe that the interface used on Facebook currently produces more of these the models that support the suggestion of the label (eg del.icio.us).

Representing the ontology that is as graph of concepts that have multiple tags can be associated variant spellings of a probabilistic way. The most common spelling is labeled natural.

B. Exploring morphological tools.

It also explores the morphological analyzes, although are focusing on the potential of combining aspects. Initial analysis of the data indicates that certain morphological techniques (Eg, remove from verb gerund plural) may be appropriate to some aspects and not to other. A significant problem with the assumption is in its common use, people tend to name generic concepts (or too general or too specific). In particular, use some specific generic concepts such as "country" or "continent" for localization, and "mammal" or "plant" for an image. In the results, for example, certain country names were specified and thus positions rarely below cities. However, these ontological concepts above are freely available in the form of gazetteers and common taxonomies. It plans to enter our new model with these ontologies model over a specific domain (DUMOs). This address is weaknesses inherent in the assumption, serving another purpose as well. Specifying the level structure upper ontology, we can enforce the model aspect that makes the most sense for users; since it is an entry in the model can be tested variants in this easily with the user base.

C. Moderation support community

While waiting for the refined model reduces the noise (Errors) in our results, it is believed that the model can deploying improved as a process not fully automated, but rather as a productivity tool. Many applications have a model labeling established for the community, including enthusiast's popular domains moderators to side. If the statistical model may suggest ontology to a set advisors, need only approve or reject the proposed relationships. Once a line is established depth, requires little effort to keep advisers fresh current ontology, reflecting current usage. The model reflects statistical community use, with moderators acting as supervisors and to balance.

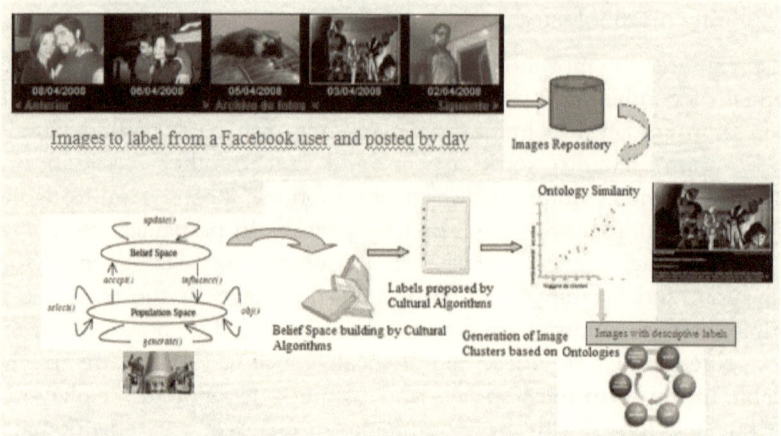

Figure 2. Proposed Methodology using Cutural Algorthms to decide the correct labeling of pictures

VI Conclusions

Described a model based on assumptions to induce ontologies using the label that produces results promising initial results. Refine this model we expect improve the accuracy, and also induce ontological. The results support more effective interfaces Search and browser, and can reasonably integrate existing models of community moderators enthusiasts as balancers

Acknowledge

At sample of students from Industrial Design at Juarez City University whom select the pictures more according at place and the context where was located these.

References

[1] Clough, P., Joho, H. and Sanderson, M., Automatically Organizing Images using Concept Hierarchies, Proc. In: Proceedings of Multimedia Information Retrieval 2005.

[2] Dumais, S, et al., Stuff I ve seen: A system for personal information retrieval and re-use. In SIGIR, 2003

[3] Hearst, M., Automatic Acquisition of Hyponyms from Large Text Corporation, in "Proc. of COLING 92", Nantes.

[4] Hearst, M User Interfaces and Visualization. In: Baeza-Yates, R. Ribeiro-Neto, B. (eds.), Modern Information Retrieval, pp. 257-323. New York: ACM Press.

[5] Mani, I., Samuel, K., Concepcion, K., and Vogel, D. Automatically Inducing Ontologies from Corpora. Proceedings of CompuTerm 2004: 3rd International Workshop on Computational Terminology, COLINGO 2004, Geneva

[6] Naaman, M, et al. Context Data in Geo-Referenced Digital Photo Collections. In proceedings, Twelfth ACM International Conference on Multimedia (ACM MM 2004), October 2004.

[7] Sanderson, M. and Croft, B. Deriving concept hierarchies from text In: Proceedings of the 22nd ACM Conference of the Special Interest Group in Information Retrieval;1999, pp. 206-213.

[8] Yee, K-P., Swearingen, K., and Hearst, M. Faceted metadata for image search and browsing. In: Proceedings of the SIGCHI conference on Human factors in computing systems; 2003, pp. 401-408.

Identification of popularity to labeling a map-based on social networking using an Imaginary Collective and Cultural Algorithms

Daniel Azpeitia[1], Alberto Ochoa-Zezzatti[1], Rubén Jaramillo[2] & Sandra Bustilos[1]

[1]Juarez City University, México

[2]LAPEM, México

alberto.ochoa@uacj.mx

Abstract. In the present study, we describe some promising initial results from induced ontology vocabularies of labels-which a map associated with a Social Networking online were selected using Cultural Algorithms, using a model based on assumptions. Describes the utility of ontology of aspects as a supplement to a system that labeled and marked with the model and show results. Proposing a revised model using probabilistic ontologies seed to induce ontology of aspects, and describe how the model can be integrated into the logistics of labeling communities.

Keywords: Cultural Algorithms, Pattern Recognition and Decision Support System.

II.   Introduction

In recent years has seen rapid growth in the Use of labeling applications, both in the number of labeling applications associated with images or maps and in the number of users participating in communities labeling. This growth currently exceeds our understanding of how entries are made which are efficient and productive for a range of applications and users. The labeling schemes are often placed in opposition to the taxonomic models and two types are they are commonly cited:

1. User interfaces for annotation based on a closed vocabulary, hierarchical are clumsy and inflexible

2. A strict tree of concepts does not reflect its use and intention.

The first criticism is valid, but can be treated easily with dynamic ontologies better mechanisms for UI. Much of the second critical not so much an issue with the taxonomy (ontology) alone itself, but something problematic which force models users to put concepts into a single hierarchy. Much of this issue can be treated using ontologies aspects that separate the various aspects of the attempt annotation. Some of the common facet used in the annotation of the media include the location, activity or event associated with several aspects of the paint (People, plants, animals, objects, in anothers.) and especially in shared context, given the emotional response. Labels provide a simple and direct to create annotations that reflect a variety of aspects, and also provide direct means of shipping on a search. The search at least based purely labels tend to have low memory operation (This can be mitigated in part by a UI that adapts a vocabulary aligned). Moreover, when a search Initial returns a large number of results, the tags not support efficient or intuitive models of refinement of the question. In the best case, users currently can refine a search using clusters of (Statistical) concepts. Although it is sometimes useful, clustering performance that is very difficult to assess. In [7] described the distinction between polythetic clusters (in which members of the cluster share a certain proportion of system characteristics) and monothetic clusters in which all members share a characteristic in common. Because of this in [7] discussed that users can understand more easily the monothetic clusters.Furthermore, because this polythetic clusters are difficult labeling, compared with monothetic clusters that are easy to label adapting to different paradigms common interface such as guided navigation, hierarchies limited to the refinement of the question, among other. In [2] have been explored search engines based on issues, as well as in [4] and on the other hand [8] show the usefulness of the search interfaces aspects to search for images. Although some in the community marking with labels resist the taxonomy, del.icio.us even (http://del.icio.us) recently added "Packages" - while it is purely organization (the packages do not support the ontological semantics), the characteristic recognizes the challenges of scaling a model organization marking with labels. It is believed that users should not having to choose between models based purely label models with purely taxonomic vocabularies closed. We explore a model that balances technical natural language processing statistics together with domain knowledge ontology to induce can be balanced on the final answer.

The objective is a system that maintains the flexibility of the interface tagging for annotation while also benefits from the power and usefulness of an ontology aspect in the search and display interface. We present initial results of a model based on symbolic logic to the labeling system as the maps associated with Facebook. (http://www.facebook.com/profile.php?id=829140429&sk=map), demonstrating the potential for induces the technique suitable for an ontology search and display the user interface. The remainder of paper describes the approach, the test set using cultural algorithms and evaluation, coupled with a proposal for a refined model and how it fit into the logistics of a brand community by tags like occurs on Facebook

II Related Research

In [7] described a simple statistical model of logic symbolic which assumes X to Y if:

$$P(X \mid Y), \ 0.8 \text{and} \ P(Y \mid X) \leq 1$$
(1)

To do this in [7] applies this model of co-occurrence the terms of the concept of the documents extracted recovered to a question addressed (use a search "query" is a great help to align the domain of terms). In [1] adapted the technique to assist (ie, recorded professionally) photographs of a historical collection. The resulting taxonomies are quite noisy (ie, many of the proposed pairs the assumption is incorrect), especially given that domain vocabularies are focused by questions original. The results are included in the table below as a baseline. Despite not working properly, these models generate the taxonomy that reflects the use real, and thus it adequately to labeling applications. Many other research have experimented with using techniques to induce ontology PLN statistics including [3, 5 and 8]. a few of These [3 and 5] are dependent at least in part of speech grammar, because it is that it can only be applied on natural language contexts. Other [8] tries to match concepts to existing ontologies such as WordNet; these models may be inherently less noisy but since WordNet is based on English vocabulary standard, this can be difficult to adapt these models to dynamic and sometimes idiosyncratic vocabulary that emerges in labeling applications (eg to name events).

III. Exploratory Approach

A. Passing Fancy.

We adapted the model proposed in [7] labeling system using on Facebook, adjusting statistical thresholds for reflect the ad hoc use, and adding filters to control for highly idiosyncratic vocabulary. So X potentially including if:

$$P (X \mid Y) \geq t \text{ and } P (Y \mid X) < t$$
$$D_x \geq D_{min}, D_y \geq D_{min}$$
$$U_x \geq D_{min}, U_y \geq D_{min}$$

(2)

Wherein: t is the tendency for co-occurrence, D_x is the number of documents in which the term x occurs, and may be greater than a minimum value D_{min}, and is U_x number of users that use x in at least one entry image, and may be larger than a minimum value U_{min}. It filters the input documents (ie photos) requiring a minimum of two terms of the label to that co-occurrence was defined. A series experiments, varying the parameters t, D_{min}, and U_{min}. We sought a balance that minimized the error rate and maximized the number of pairs proposed assumption. Using more stringent values for the threshold of co-occurrence (Approaching 0.9) reduces the error rate substantially but dramatically reduces the number of pairs proposed. Useful values were between 0.7 and 0.8, below the comparable were determined empirically as in [7]. The model is more sensitive to changes in U_{min}, which D_{min}. U_{min} set to any value below 5 occurred many highly idiosyncratic in terms noisy pairs of assumption, a useful range was 5 to 20. Values D_{min} varying from 5 to 40, and proved useful as through tuning the value. Both values were increased slowly while the number of documents was increased. With the fixed input below 1 million photos, vocabulary was less stable and thus the model was sensitive to the parameters. B. Tree pruning and reinforcement.

Once you calculate the statistical co-occurrence, pairs of candidate terms are selected using the specified constraints. Then a graph is built possible relationships of father-son, and filtered out the co-occurrence with the ancestors of nodes that are logically about his father. Once the statistic is calculated the co-occurrence, selected pairs of the term of candidate using the specified constraints. Subsequently, builds a graph of the possible relationships of father-child,and filter out the co-occurrence of nodes with ancestors that are logically about his father. Ie for a given ratio of the term should be strengthened. We increased the weights of each accordingly. Finally, with each leaf on the tree you choose the best path to a root, since the weights (reinforced) of the co-occurrence for prospective parents for each node, and we join paths in trees. With systems of document sufficiently large, many of the tres result are quite large - example, cities with points of interest. There was a disproportionate number wrong paths in single-instance substructures (Singleton) and double instance (doubleton) with respect to the largest sub-structures, and filter these out jointly. This is justified since the number candidate's total tree was too big for these runs (from 500 to over 3000 candidate pairs meet with a basic assumption and the criterion of filtration), and since the ultimate goal is to provide enough structure to assist in making sense navigation and guidance through the collection. A goal improve secondary search terms by subtracting the father for images with terms of the child, and in this sense is sacrificed some recoveries certainly filtering out trees singleton and doubleton. It thinks that users will fancy treesmore sensitive to the recovery accuracy, but this aspect of the model should be evaluated in user studieslarge scale.

IV. Dataset and their analysis.

We used a snapshot of the meta-data user from Facebook with 387 places located with a respectivaly photo or more (see Figure 1). To this date, Facebook had a total of 87 million images labeled to this date, and about 37 million entries in total. About 5 million of these images were scored as "not public", so were excluded from the experimental system. The tables were modified by anonymizing for all user data (including IDs photo) and all images with less than 2 terms were filtered. This resulted in a test set of about of 7000 images. The associated vocabulary was limited to 200K and generated 5000 pairs in total (no exact number is available as we filter out some while reducing Cultural Algorithms using). Using this Evolutionary Computation technical aspects determine cultural community to assess. Among the annotations Facebook, vocabulary is inconsistent with respect to limits spelling and word (eg "Los Angeles" shows often as they can be analyzed as two terms "The" and "angels", due to some non-intuitive interface input label). In addition, many terms idiosyncratic annotation. These latter vary personal events described as a phrase label ("johnandmaryswedding" - possibly indicating some confusion.

A. Evaluation of the assumption Trees that were evaluated were manually. Each pair was marked assumption proposed as correct, inverted, related, synonymous (variants including the language in common terms as flower "/" blume "/" fleur "/" bloem "among others), or noise (entirely error). The Figure 2 shows the functionally of this research. Many of the concepts the child "Los Angeles" are suburban and points of interest; several are (possibly) related and is an example of entropy result of a statistical model. In the second example, each child node is a hyponym of "Crystal" although perhaps an art historian would create as domain model "right" in the representation use within the community from Facebook. Based on our experience and that of others (eg [6]) is presumes that the images will be recorded and retrieved at more easily possible to emphasize various aspects of the keyword: place, activity, and images. The community of Facebook also another aspect that seems to accentuate might be described best as emotion or response. In this paper the results of a proportion large shared vocabulary is tied to the names of places, although it counted with the refinements of the modelto produce more balance with other aspects. For location are considered a combination of names geographical locations as well as points of interest demarcate the most to the activity. Thus we consider "Los Angeles" reasonable parent "Chinese Theater". In the sense of a pure type of relationship does not hold, however for the utility to locate an image, entirely reasonable. In the same way "Los Angeles" may be related to, but not a parent of "muni" or "Streetfair". For generic terms such as "lake" and "park" considered instances of lakes or parks that could be reasonably children. In the images, more typical of the type relations were used: "dog" but included the specific races, "food" "kimchee" and "Creamcheese" where"Restaurant" is only related. In a large photo that shares the environment such as Facebook, relationships are less useful personal question, so it considered almost all personal names such as noise any context of the pair. Table I compares theresults related models for the assumption results.

Figure 1. View a personal site with map from Facebook.

TABLE I

COMPARISON OF RESULTS

Model	Average error/another	Relations	Correct	Relation/Aspect	Equals
A	? 19%	23%	49%	8%	
B	105 0.2%	15%	10%	43%	
C	1200 +	51%	21%	5%	23%

A Sanderson and Croft7.

B Clough et al1.

C Proposed in this research..

In [7] reported a high "aspect of" numbers, and attributed this at least in part due to questions vocabulary-limiting. In [1] presents an application similar to ours and so was given a line useful background. it is believed that his model would be made better if was applied to the entire whole vocabulary of data rather than a focused question. Both in [1 and 7] appears to contain an inconsistency in the statistical model (The second term should be expressed as $P\ (X\ |\ Y)\ <0.8$ and not $P\ (X\ |\ Y)\ <1$, although this may be a typo in the articles.

V. Proposed Model

Initial results are very promising and give rise to additional work. The proposed model produces subtrees that generally reflect different aspects but cannot categorize concepts aspects. It has planned a series of changes to the model to address this. A. Migrating to a pure probabilistic model

Currently research to express the assumption, pruning the tree construction, and classification of appearance together in a unified probabilistic model, something similar to that proposed in [5]. A probabilistic model should be more robust, and incorporate concepts such as "the number of authors using a label "as property ladder rather than a simple threshold as in the current model. Also want to add better support for replays and spelling errors, we believe that the interface used on Facebook currently produces more of these the models that support the suggestion of the label (eg del.icio.us).

Representing the ontology that is as graph of concepts that have multiple tags can be associated variant spellings of a probabilistic way. The most common spelling is labeled natural.

B. Exploring morphological tools.

It also explores the morphological analyzes, although are focusing on the potential of combining aspects. Initial analysis of the data indicates that certain morphological techniques (Eg, remove from verb gerund plural) may be appropriate to some aspects and not to other. A significant problem with the assumption is in its common use, people tend to name generic concepts (or too general or too specific). In particular, use some specific generic concepts such as "country" or "continent" for localization, and "mammal" or "plant" for an image. In the results, for example, certain country names were specified and thus positions rarely below cities. However, these ontological concepts above are freely available in the form of gazetteers and common taxonomies. It plans to enter our new model with these ontologies model over a specific domain (DUMOs). This address is weaknesses inherent in the assumption, serving another purpose as well. Specifying the level structure upper ontology, we can enforce the model aspect that makes the most sense for users; since it is an entry in the model can be tested variants in this easily with the user base.

C. Moderation support community

While waiting for the refined model reduces the noise (Errors) in our results, it is believed that the model can deploying improved as a process not fully automated, but rather as a productivity tool. Many applications have a model labeling established for the community, including enthusiast's popular domains moderators to side. If the statistical model may suggest ontology to a set advisors, need only approve or reject the proposed relationships. Once a line is established depth, requires little effort to keep advisers fresh current ontology, reflecting current usage. The model reflects statistical community use, with moderators acting as supervisors and to balance.

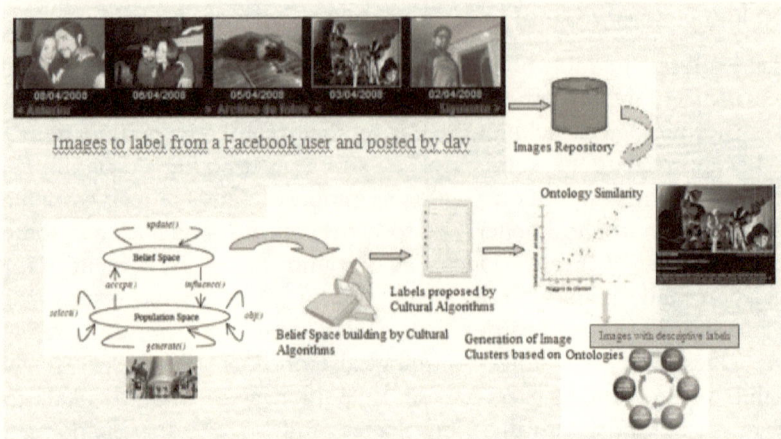

Figure 2. Proposed Methodology using Cutural Algorthms to decide the correct labeling of pictures

VI Conclusions

Described a model based on assumptions to induce ontologies using the label that produces results promising initial results. Refine this model we expect improve the accuracy, and also induce ontological. The results support more effective interfaces Search and browser, and can reasonably integrate existing models of community moderators enthusiasts as balancers

Acknowledge

At sample of students from Industrial Design at Juarez City University whom select the pictures more according at place and the context where was located these.

References

[1] Clough, P., Joho, H. and Sanderson, M., Automatically Organizing Images using Concept Hierarchies, Proc. In: Proceedings of Multimedia Information Retrieval 2005.

[2] Dumais, S, et al., Stuff I ve seen: A system for personal information retrieval and re-use. In SIGIR, 2003

[3] Hearst, M., Automatic Acquisition of Hyponyms from Large Text Corporation, in "Proc. of COLING 92", Nantes.

[4] Hearst, M User Interfaces and Visualization. In: Baeza-Yates, R. Ribeiro-Neto, B. (eds.), Modern Information Retrieval, pp. 257-323. New York: ACM Press.

[5] Mani, I., Samuel, K., Concepcion, K., and Vogel, D. Automatically Inducing Ontologies from Corpora. Proceedings of CompuTerm 2004: 3rd International Workshop on Computational Terminology, COLINGO 2004, Geneva

[6] Naaman, M, et al. Context Data in Geo-Referenced Digital Photo Collections. In proceedings, Twelfth ACM International Conference on Multimedia (ACM MM 2004), October 2004.

[7] Sanderson, M. and Croft, B. Deriving concept hierarchies from text In: Proceedings of the 22nd ACM Conference of the Special Interest Group in Information Retrieval;1999, pp. 206-213.

[8] Yee, K-P., Swearingen, K., and Hearst, M. Faceted metadata for image search and browsing. In: Proceedings of the SIGCHI conference on Human factors in computing systems; 2003, pp. 401-408.

Intelligent optimization to small spaces using IKEA furniture

Alberto Ochoa-Zezzatti[1], Ruben Jaramillo[2], Sandra Bustillos[1], Nemesio Castllo[1], & Manuel García-Cossío

[1] Juarez City University, México

[2] LAPEM, México

[3] UNINTER, México

alberto.ochoa@uacj.mx

Abstract. The paper discusses a researh related with the innovative sense of using Decision Support System based on a Bioinspired Algorithm related with optiizate the space of a 25m^2 associated with a real problem of space in our lives, to determine the correct and adequate selection of furniture in an standard apartment with small spaces, and location of diverse issues to analyze the way to improve the home of many families in Mexico, this research which permits select a specific number of issues from the catalogue of IKEA, these furniture issues are evaluated from a repository with data from another suceesful accomodations. Each issue was analyzed to built their cost-benefit during different scenarios and determine the viability of using specific furniture issues using a formal methodology based on Bioinspired Algorithms. An IKEA Catalogue is characterized and analyzed by obtain the most representative and sucessful future scenario to determine the quantity of issues to utilize which try to improve the limited resources and the perspectives of determine the correct selection to stablishment a confortable home. A case of study is presented using a limited space. In addition, we analyzed the selection and location of diverse issues using a similarity model to locate this. The sample of study allowed analyzing the individual features of each issue with the emulation from set matching features.

Keywords: Cultural Algorithms, Pattern Recognition and Decision Support System.

1. Introduction

IKEA is a privately held, international home products company that designs and sells ready-to-assemble furniture such as beds, chairs, and desks, appliances and home accessories. The company is the world's largest furniture retailer founded in 1943 by 17-year-old Ingvar Kamprad in Sweden, the company is named as an acronym comprising the initials of the founder's name (Ingvar Kamprad), the farm where he grew up (Elmtaryd), and his hometown (Agunnaryd, in Småland, South Sweden).

The firm is known for the attention it gives to cost control, operational details and continuous product development, allowing it to lower its prices by an average of two to three per cent over the decade to 2010, while continuing its global expansion.

The groups of companies that form IKEA are all controlled by INGKA Holding, a Dutch corporation, which in turn is controlled by a tax-exempt, not-for-profit Dutch foundation. The IKEA trademark and concept is controlled by a series of corporations that can be traced to the Netherlands Antilles and to the Interogo Foudation in Liechtenstein.

INGKA Holding B.V. owns the industrial group Swedwood, which sources the manufacturing of IKEA furniture, the sales companies that run IKEA stores, as well as purchasing and supply functions, and IKEA of Sweden, which is responsible for the design and development of products in the IKEA range. INGKA Holding B.V. is wholly owned by Stichting INGKA Foundation, which is a non-profit foundation, registered in Delft, Netherlands. The European logistics centre is located in Dortmund, Germany, and the Asian logistics centre is located in Singapore.

Inter IKEA Systems B.V. in Delft, also in the Netherlands, owns the IKEA concept and trademark, and there is a franchising agreement with every IKEA store in the world. The INGKA Group (not to be confused with INGKA Holding B.V.) is the biggest franchisee of Inter IKEA Systems B.V. Inter IKEA Systems B.V. is not owned by INGKA Holding B.V., but by Inter IKEA Holding S.A. registered in Luxembourg which in turn is controlled by the Interogo Foundation in Liechtenstein. Ingvar Kamprad has confirmed that this foundation is controlled by him and his family. The company which originated in Småland, Sweden, distributes its products through its retail outlets. As of October 2011, IKEA has 332 stores in thirty-eight countries. In fiscal year 2010, it sold $23.1 billion worth of goods, a 7.7 percent increase over 2009. On February 17, 2011, IKEA announced its plans for a wind farm in Dalarna County, Sweden, furthering the furniture giant's goal of running on 100 percent renewable energy.

The IKEA website contains about 12,000 products and is the closest representation of the entire IKEA range. There were over 470 million visitors to the IKEA websites in the year from September 2007 to

September 2008. IKEA is the world's third-largest consumer of wood, behind The Home Depot and Lowe's.

2. Cultural Algorithms

The initial development of Cultural Algorithms (CAs) can be attributed to Reynolds [12] this approach is a complement to the metaphor used by evolutionary algorithms, which had focused on the concepts of genetics and natural evolution. Cultural algorithms are based on the theories of anthropologists, sociologists and archaeologists, who have tried to model the evolution as a process of cultural evolution [5]. The belief space characterizes CAs as evolutionary algorithms, which are used to store the acquired knowledge from previous generations. The information in this space must be accessible to any individual, who may use it to change their behavior and their respective proposed solution. To join the belief space and the population is necessary to establish a communication protocol, which dictates rules of the type of information to be exchanged between spaces. This protocol defines the acceptance and influence functions. The acceptance function is responsible for accepting the information or the experience that individuals have obtained in the current generation and transport into the belief space. On the other hand, the influence function is responsible for "influencing" variation operators (e.g. crossover and mutation in the case of genetic algorithms). This means that this function set some kind of pressure on resultant individuals from the application of variation operators to reach the desirable behavior, also away from undesirable results, always according to information stored in the belief space.

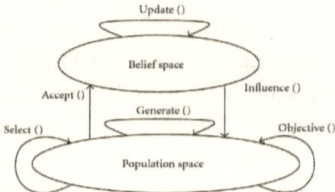

Figure 1 Different Spaces employed by the Cultural Algorithm.

Figure 1 presents the interaction between the belief space and population space. The population space works similar to that of an evolutionary algorithm, i.e. the population consists of a set of individuals where each has an independent feature used to determine their suitability (fitness). The interaction between the two spaces makes the cultural algorithm increases the complexity in the development and computation of the evolutionary algorithm. Below we show the general pseudo-code of a cultural algorithm.

```
Begin
  t=0;
  Initialize POP(t); // Initialization of population
  Initialize BLF(t); // Initialization of believing space
  Evaluate POP(t);
  While (Do not condition of term t=t+1)
    Vote (BLF (t), Accept (POP(t))));
    Adjust (BLF (t));
    Evolve(POP(t), Influence(BLF(t)));
    t = t +1;
    Select POP(t) from POP(t-1);
  End While
End
```

3. Selection of adequate furniture to small apartments.

Design of small spaces require of many features to determine the quantity, distributions of issues and their location to improve the components of each space which is analyzed in the diverse IKEA catalogues as the shown in the Figure 2.

Figure 2.- Ikea Catalogues with issues in different spaces.

We try to improve the next equation 1 using the table 1, in which is V as the value of the property and define with different aspects relative with the potential features of the house.

$$V = CHU - [\#R * SH]^{RA} \pm SCAH$$

(1)

$$\qquad\qquad IHI \qquad PLH$$

In where:

CHU = Cost of House Unit.

IHI = Increase House Index

#R = Number of rooms in the house.

SH = Size of House.

PLA = People living on the house.

RA = Recreational Areas n the house.

SCAH = Social Cultural Aspects related to the location of issues on the House.

4.   Multiple Matching.

The multiple matching is a series of seven evaluations according to different combinations of furniture issues and a batch of 50 runs under different scenarios. In the evaluation phase economics specifications with more similarities will be given a preference, and then these aspects will be selected to compete. Furniture issues must be ranked according to their customers' preferences after tournaments end once the final list of multiple matching is evaluated. The hybrid algorithm sets the right for customers to evaluate a batch according to the organizational needs and the issues for each comparison assign the issues list before a new cycle begins. The hybrid algorithm will be scheduled to set the timing for the comparison of different similarities using a round of multiple matching analyses based in the commercialization assigned to a issue. Then, issues that qualify for selection in a Model will be chosen on the following prioritized basis. For the first cycle of similarity, all issues in the Repository (ie. Loveseat or Lamp) will be invited to participate for different comparisons. Given the organization for each issue and the matches for each round in the algorithm, issues are asked to state their participation for its evaluation in each of the series. In case any of these issues decline to participate in the series, the algorithm may nominate one issue to be set as a replacement, and this issue has to be rated amongst the top issues in the Repository. Based on an average calculation of two decimal places, the rating list in the series of comparisons, before starting a new cycle, ten qualifiers will be selected (excluding the seven issues that will be compared in the matches). In case issues have the same average rating, the number of similarities set for the match will be used to determine its ranking. To ensure an active participation in the future, a minimum of twenty-five games are recommended for the four included rating lists and before the main rating list. When a issue does not accept to play into a Multiple Matching series, then the selection process uses the average rating plus number of games played during the rating period. The algorithm repeats this process until reaching the required qualifiers of the Multiple Matching series.

4   Experimentation.

In order to obtain the most efficient arrangement of issues, we developed a cluster for storing the data of each of the representative individuals for each issue. The narrative guide is made with the purpose of distributing an optimal form for each evaluated issues [9]. The main experiment consisted in implementing issues in the Cultural Algorithm, with 500 agents and 200 beliefs into the belief space. The stop condition is reached after 50 runs; this allowed generating the best selection of each kind and their possible location in a specific Model. A location is obtained after comparing the different cultural (color, material, form) and economical similarities of each issue and the evaluation of the Multiple Matching Model as in [10]. The vector of weights employed for the fitness function is $W_i=[0.6, 0.7, 0.8, 0.5, 0.6]$, which respectively represents the importance of the particular attribute: Cost, Size, Cares associated with the furniture, Ergonomic and Cost-Benefit. Then, the cultural algorithm will select the specific location of each issue based on the attributes similarity. Each attribute is represented by a discrete value from 0 to 5, where 0 means absence and 5 the highest value of the attribute. The experiment design consists of an orthogonal array test with interactions amongst the attribute variables; these variables are studied within a location range (1 to 400) specific to a coordinates x and y. The orthogonal array is $L-N(2**5)$, in other words, 5 times the N executions. The value of N is defined by the combination of the 5 possible values of the variables, also the values in the location range. In Table 1 we list some possible scenarios as the result of combining the values of the attributes and the specific location to represent a specific issue. The results permit us to analyze the effect of the variables in the location selection of all the possible combinations of values.

Table 1 The orthogonal array test.

Cost	Size	Carefully	Ergonomic	Cost-Benefit
4	1	2	2	3
3	1	2	2	3
2	1	3	2	4
5	1	3	2	5

The use of the orthogonal array test facilitates the reorganization of the different attributes. Also the array aids to specify the best possibilities to adequate correct solutions (locations) for each issue. Different attributes were used to identify the real possibilities of improving a issue set in a particular environment, and to specify the correlations with other issues (see Figure 3). The locations will be chose based on the orthogonal test array.

Figure 3.- Location of issues in a most high evaluated scenario according to spaces.

5   Conclusions.

These patterns represent a unique form of adaptive behavior that solves a computational problem that does not make clusters of furniture issues. The resultant configurations can be metaphorically related to the knowledge of the behavior of the community with respect to an optimization problem (to culturally select 25 similar issues [4]). Our implementation related each of the issues. The Narrative guide, allowed us to identify changes in time related associated with issues. Therefore, we realize that the concept of "negotiation" exists based on determining the acceptance function to propose an alternative location for the rest of issues [8]. Additionally, this may help to understand true similarities that share different issues based in the characteristics to be clustered and also to keep their own design. In a related work [7], it has been demonstrated that small variations are developed through the time. On the other hand, CAs can be used in the Evolutionary Robotic field where social interaction and decision is needed, for example in the training phase described in [11]. Finally, CAs can be used in pattern recognition in a social database, for example: fashion styling and criminal behavior.

References

23.   Desmond, A. and Moore J. (1995) Darwin – la vida de un evolucionista atormentado. Generación Editorial, São Paulo, Brazil
24.   Ochoa A. et al. (2007) Baharastar – Simulador de Algoritmos Culturales para la Minería de Datos Social. In Proceedings of COMCEV'2007
25.   Ochoa, Alberto et al. (2009) Dyoram's Representation Using a Mosaic Image, The International Journal of Virtual Reality
26.   Memory Alpha, http://en.memory-alpha.org
27.   Callogerodóttir Z. and Ochoa A. (2007) Optimization Problem Solving using Predator/Prey Games and Cultural Algorithms NDAM'2003, Reykiavik; Iceland
28.   Tang Hué et al. (2006) The Emergence of Social Network Hierarchy Using Cultural Algorithms, VLDB'06, Seoul, Korea
29.   Vukčević I. and Ochoa A. (2005) Similar cultural relationships in Montenegro JASSS'2005, England
30.   Zuckermann Dennis (1991) Culture and Organizations, London: McGraw-Hill,
31.   Ponce J. et al. (2009) Data Mining and Knowledge Discovery in Real Life Applications", Book edited by: Julio Ponce and Adem Karahoca, ISBN 978-3-902613-53-0, pp. 438, I-Tech, Vienna, Austria
32.   Ustaoglu, Yesim. (2009) Simulating the behavior of a minority in Turkish Society ASNA'2009, Zurigo, Svizerra
33.   Nolfi S. and Floreano D. (2000) Evolutionary Robotic: The Biology, Intelligence, and Technology of Self-Organization Machines, MA: MIT Press.
34.   Robert G. Reynolds and William Sverdlik. (1994) Problem Solving Using Cultural Algorithms. International Conference on Evolutionary Computation, 645-650.

A multivariate analysis of mobbing and their organizational implications

Alberto Ochoa[1,*], Sandra Bustillos[3], Petra Salazar[1], and Nemesio Castillo[1]

[1] Juarez City University, México

alberto.ochoa@uacj.mx

Cisneros scale to assess psychological harassment or mobbing at work. The Cisneros scale is a selfmanaged questionnaire developed by Iñaki Piñuel that contains 43 items in a seven-point Likert response scale. The range is from never (0) to daily (6). Items deal with 43 different behaviours related to bullying, harassment and emotional abuse in the workplace. The scale forms part of a more extensive instrument, the CISNEROS Barometer ™ (*Cuestionario Individual sobre PSicoterror, Negación, Estigmatización y Rechazo en Organizaciones Sociales*; Individual Questionnaire on Psychoterror, Mobbing, Stigmatization and Rejection in Social Organizations), which includes 3 scales for exploring periodically the content and development of violence and harassment in the workplace in Spain. In this paper we analyze the reliability, validity and dimensionality of the Cisneros scale using data from a sample of 1.303 workers from different sectors. Results show that Cisneros has very high reliability (0.96) and a bidimensional structure that includes «Locus of bullying behaviour» and «Type of bullying behaviour», and that it presents expected and statistically significant correlations with scales measuring Self-esteem, Burnout, Depression, Intention to leave, and Post-traumatic Stress. It is concluded that Cisneros is capable of discriminating between different levels of bullying and harassment at work and can be considered a valuable tool to examine and establish the frequency and damage of mobbing at organizational or individual level.

Introduction.

Mobbing causes enormous suffering to the people who suffer and decrease the potential competitiveness of the organizations in which they occur (for review see Einersen Zapf, 2001). In recent years there have been several definitions of mobbing. Among the most important are:

Chain over a fairly short period of attempted or consummated hostile actions, expressed or expressed, by a person or persons, to a third: the goal. Mobbing is a destructive process, consists of a series of hostile actions, which taken in isolation, may seem harmless but whose constant repetition has pernicious effects (Leymann, 1996a).

In the workplace, we define bullying as verbal abuse continued and deliberate and modal a worker receives from another or others, who treat him cruelly in order to achieve psychological annihilation or destruction and to obtain his release from organization through different procedures illegal, illicit, or outside or humanitarian respectful treatment and undermine the dignity of the worker. Mobbing aims to intimidate, emotionally and intellectually to the victim in order to remove it from the organization or satisfy the insatiable need to attack, control and destroy it usually has the harasser, who takes the opportunity that gives you the particular organizational situation (reorganization, cost reduction, bureaucratization, rapid change, among others) to push a number of pulses and psychopathic tendencies (Piñuel, 2001a).

All abusive behavior (gestures, words, behaviors, attitudes ...) that attempts by their repetition and consistency to the dignity or psychological or physical integrity of a person, jeopardizing their jobs or degrading work climate (Hirigoyen, 2001).

In all these definitions are two characteristic features of mobbing, a temporal persistence of bullying behavior and its wide range of possibilities. The strategies used to subject the victim to bullying are varied, and most often are combined with one another as a "comprehensive treatment" (Piñuel, 2001a). Etre These include the following:

-Shouting, screaming, pushy or insulting the victim when alone or in the presence of others.

-Assign projects with deadlines or targets that are known unattainable or impossible to meet, and endless tasks that are clearly at the time.

-Take away key areas of responsibility, routine offering in return, without interest or even no work to perform ("Until I get bored and go").

-Ignore, exclude, speaking only to a third person present, simulating its nonexistence (ninguneándolo) or no physical presence in the office or in meetings he attends ("like invisible").

-Withholding information crucial to their work or to manipulate it to mislead in their job performance, and then accuse him of negligence or professional misconduct.

Company-Extender for malicious or slanderous rumors that undermine their reputation, their image or their professionalism.

-Underestimating or not at all appreciate the effort made by the refusal to recognize a job well done or waiver periodically evaluate their work.

-Ignore the professional success or maliciously attributing them to others or elements outside it, as chance, luck, market conditions, among others.

-Continually criticize their work, their ideas, proposals, and solutions, among other featuring in a way to anime cartoon or parodying.

Punishing hard-decision or taking any personal initiative in the performance of work as a serious breach of duty of obedience to the hierarchy.

-Ridiculing his work, his ideas and the results obtained with other workers.

-Encourage other students to participate in any of the above actions by persuasion, coercion or abuse of authority.

Invading the privacy-speaking harassed your mail, phone, checking their documents, cupboards, drawers etc., maliciously subtracting key elements to his work.

The result that the behaviors described have labor welfare and physical and psychological health of those suffering are easily predictable, including: depression, anxiety, posttraumatic stress, "burnout" and professional neglect (Piñuel, 2003). It is considered the Swedish psychologist Heinz Leymann, the first researcher who systematized and widely distributed (1986, 1990, 1993, 1996a) the phenomenon known as mobbing and its consequences, though it had been defined previously by Lorenz (1991) in ethology and Heinemann (1972) in education.

The book edited by Leyman (1993) in German, followed by other reference works in French (Hirigoyen, 1994), English (Davenport, Schwartz and Elliot, 1999) and Spanish (Piñuel, 2001, 2003). A Leyman (1992) must be also the first questionnaire designed to assess the mobbing, the LIPT (Leyman terrorization of Psychological Inventory). The LIPT (adapted to Castilian by Gonzalez de Rivera and Rodriguez-Abuín, 2003) is a questionnaire describing 45 bullying behaviors and should indicate whether the person has been assessed or not. In 2000 Prof. Iñaki Piñuel, in order to assess not only the mobbing at work in organizations, but also its consequences,

CISNEROS developed the barometer, the first tool used to measure the incidence of the phenomenon of bullying in Spain. This questionnaire has a scale specifically designed to assess bullying behavior. This scale, called Cisneros scale, following the same guidelines as the LIPT, objective 43 bullying behaviors requesting the respondent that values on a scale of 0 (never) to 6 (every day) the degree to which it affected for each of the harassing behavior. Bookmark the barometer CISNEROS was first applied in 2001 for active workers of all sectors of the Community of Madrid in the vicinity of Alcala de Henares and Guadalajara. It was the first monographic research with a large sample in Spain with almost 1,000 valid questionnaires. The research results were published by the Journal AEDIPE, winning the second prize at the award for best scientific paper in HR 2001 (Piñuel, 2001b). This first research have followed others, some of whose results are already published (Piñuel and Oñate, 2002, 2003), or will be released soon (Piñuel, Fidalgo, Oñate and Ferreres, 2004a, 2004b).

In this context, the objective of this research is to determine the reliability and validity of the scale as a screening instrument Cisneros of bullying at work. Furthermore, in the area of assessing the validity of the scale, we will study its relationship with syndromes such as burnout, posttraumatic stress syndrome, depression and psychological variables such as self-esteem, which, in principle, can be affected by a mobbing situation. Given the extent of the investigation will be presented only those results relevant in order to establish the properties psychometric scale.

Methodology developed.

Instrument

CISNEROS Barometer is a questionnaire developed by prof. Piñuel periodically to poll the state and the levels of violence in the workplace. It consists of three scales, Cisneros scale to assess the degree of suffering bullying, a second scale for assessing posttraumatic stress built from the diagnostic criteria of DSM-IV, and third professional Abandonment scale that measures the intention of the person to change jobs and / or careers.

Along with these scales, was applied to the sample questionnaire on Burnout MBI (Maslach and Jackson 1997), the Beck Depression Inventory in the revised version of 1979 (Beck, Rush, Shaw & Emery, 1983) and a Spanish adaptation of the Rosenberg Self-Esteem Scale (Curbow & someField, 1991, Rosenberg, 1965).

Scale Cisneros. Cisneros is objective scale and measures 43 bullying behaviors requesting the respondent that values on a scale of 0 (never) to 6 (every day) the degree to which it is affected by each of the harassing behavior. The Appendix presents a full version of the scale Cisneros.

Correction Form Cisneros scale. The Cisneros scale correction, similar to that proposed for the questionnaire LIPT-60 (González de Rivera and Rodriguez-Abuín, 2003), can be performed according to three indices.

First, the total number of strategies harassment (NEAP), is a simple count of nonzero responses, we report the total number of strategies suffered harassment. The second, the overall rate of bullying (IGAP), equal to the sum of the scores on the items divided the number of items, is a global index of the degree of harassment experienced. This index provides the same information you would get to use the total score on the test (on a scale of 0-258 points9, with the advantage that this information is now expressed in the same scale in which items are answered (from 0 to 6.) Third, the average intensity of bullying strategies (IMAP), is equal to the sum of the scores on the items divided by the index value NEAP. This index indicates the average intensity suffered bullying strategies. sense of the last two indices is clearer when considered together. The IMAP value is always greater than or equal to IGAP, as the first divides the sum of the scores on the items by the number of responses positive, while the second is always divided by the total number of items. (43) Thus, very large differences between IMAP and IGAP indicate that the subject has few harassing behavior but which are very intense experience.

Sample

The universe on which the study is the active working population in the Henares corridor area. The sample consists of 1,303 active workers over 18 belonging to companies and public bodies and training of different categories. This exhibition consists of selected passers incidentally in various neighborhoods near Alcala de Henares to which they were given the survey and returned correctly completed. The demographic and labor of the respondents are presented in Table 1.

Procedure

The procedure used for the implementation of the barometer was CISNEROS personal delivery of questionnaires, with a target several days for completion. In the heading of the questionnaire provided the following instructions: This is a research conducted by the University of Alcalá. Collaboration is required to complete the following questionnaire. Below you will find a series of questions relating to their work. This research, called Cisneros II-G, is conducted by professors at the University of Alcala de Henares under the direction of Professor Iñaki Piñuel and Zabala and aims to identify a number of problems that may be affecting the occupational health Spanish workers. The study is confidential and guarantees the anonymity of the people who filled in. It is recommended that you have a period of time in which to read quietly and record your answers to the issues presented. This may be recommended that you answer the questions in a quiet place and sufficient time. Remember that your answer will be helpful to establish and identify the impact of potential labor problems affecting workers from Spain. Thank you for your feedback.

Statistical Analysis

Features distributions

Since most of the study variables showed markedly asymmetrical distributions, has been used as an index of central tendency median (value above and below which are half of the cases), and variability as the amplitude index interquartile (difference between the values for the 75 and 25 percentile. such statistical unlike, for example, the mean and standard deviation, are insensitive to outliers, resulting not affected by a few extremely high or low values. by the same reason, non-normal distributions, was used as a correlation coefficient Spearman's rho, and to determine whether there were statistically significant differences between groups, the nonparametric equivalent of the t test, the Mann-Whitney. in all cases the normality of variables was checked using the Kolmogorov-Smirnov test. Unless expressly qu otherwise noted, the confidence level used to judge the statistical significance was 95%, in all analyzes.

Tabla 1
Descripción de las características de la muestra

	Varones	Mujeres	Total	%
Grupos de edad				
1 Menor de 23 años	161	150	311	25,4
2 Entre 24 y 30 años	216	213	429	32,9
3 Entre 31 y 40 años	115	140	255	19,6
4 Entre 41 y 50 años	105	97	202	15,5
5 Más de 51 años	71	16	87	6,7
Casos perdidos			19	1,4
Sectores de actividad				
1 Sanitario	13	56	69	5,3
2 Industria	194	90	284	21,8
3 Hostelería-Turismo	37	58	95	7,3
4 Transportes-Comunicaciones	44	16	60	4,6
5 Administración pública	63	83	146	11,2
6 Educación	17	46	63	4,8
7 Medios de comunicación	20	15	35	2,7
8 Nuevas Tecnologías	66	43	109	8,4
9 Otros	214	204	418	32
Casos perdidos			24	1,8

Reliability

The reliability of the scale was calculated by the method of the halves. The estimate of the reliability provided by this procedure is the result of applying the Spearman-Brown correction to the coefficient of correlation between the scores on the two subsets of items, chosen randomly, into which the questionnaire.

Validity

Were conducted correlational analyzes between the scale scores on scales measuring Cisneros and Self Esteem, Burnout, Depression, and PTSD Abandonment professional to see if the relationship between these variables hipoterizadas met.

Dimensionality

To evaluate the dimensionality of the data is decided using the multidimensional scaling factor analysis versus, for two reasons. First, while it is common to treat data from Likert scales as interval variables, strictly speaking, the measurement level is ordinal data, and multidimensional scaling to work easily with this data type and . Secondly, it generally scaling provides solutions mutlidimensional lower dimensionality than the factorial analysis (Real Deus, 2001). The following briefly describes the procedure followed. First, we calculated the Spearman correlation between all pairs of items (denoted by i, j), obtaining a square matrix of dimensions 43 * 43. The next step was to transform said correlation matrix at a dissimilarity matrix. They apply for the following transformation (Coxon, 1982).

$$d_{ij} = \sqrt{2\left(1 - r_{ij}\right)}$$

Being, Spearman correlation rij between items i and j. Finally, we analyze the dissimilarity matrix using a non metric scaling model using the program PROXSCAL (proximities scaling) implemented in SPSS (Version 11.0.1). As we know, multidimensional scaling to represent proximity (similarities) between the questionnaires items as distances in a low dimensional space. The objective is to find the solution with fewer dimensions allow adequately explain the data observed. To choose between the various solutions was used as an index of goodness of fit function known as Stress (S).

$$S = \sqrt{\frac{\sum_i \sum_j \left(d_{ij} - \hat{d}_{ij}\right)^2}{\sum_i \sum_j \hat{d}_{ij}^2}}$$

Where,

dij = distance between the items i and j

d ^ij = disparity between items i and j

This function informs about the gap between the distances and disparities (a transformation of the dissimilarities) we found among items. The higher the value of S, the worse the model fit to the data.

Results

Cisneros scale scores

According to the criterion of a minimum of strategies harassment of six months and a weekly (Leymann, 1996b), we divided the total sample (N = 1,303) in two subsamples, mobbing sufferers (answer yes to the question of whether they have suffered harassment behaviors for at least six months, n = 207) and those without (say no; n = 1,096). The question that is used as a decisive criterion is the number 44 (see Appendix). Table 2 shows the percentage people of each sample (total, free from harassment and harassment) who say they have suffered bullying behavior that describes each item. Parenthetically appears as the central tendency index medium, given the marked asymmetry of distributions. A simple inspection of Table 2 allows to appreciate that there a huge difference in the percentage of people who say they have each suffered bullying behavior between the sample and not harassed of harassment. Also in the average frequency of the behavior harassment: in the sample of non harassed 55.8% of the items average scores obtained from 1 (few times a year or less); beset sample in 67.4% of the items obtained a score average of 3 (sometimes a month). Consistent with these results confirmed the existence of statistically significant (Mann-Whitney, two-tailed) between bullied and harassed in the number of bullying strategies suffered (Z = -19.05, p = .00), the overall rate of bullying (Z = -20.03, p = .00) and the average rate of harassment (Z = -18.96, p = .00). Table 3 provides the average value and variability indices calculated in both subsamples harassment By contrast, no differences were found gender significant in any of the indexes Harassment: ENAP (Z = -0.19, p = .85), IGAP (Z = -0.01, p = .99) and IMAP (Z = -0.26, p = .80).

Reliability of the scale and metric properties of the questionnaire items were randomly divided into two halves, composed of 22 and 21 items, with a median of 21 and 23 and interquartile range of 10 and 13, respectively. The value of the Spearman correlation between scores on the two halves was .92, which, properly adjusted by the Spearman-Brown correction, provides an estimate of the scale reliability of .96. The estimate of the reliability provided by the Cronbach alpha coefficient was .97, confirming the high reliability of the questionnaire. Us back to Table 2 it can be seen in the discrimination indices of the items (Spearman correlation).

Tabla 2

Índice de discriminación (ID) de los ítems y porcentaje de personas que sufren cada conducta de acoso. Entre paréntesis aparece la frecuencia promedio (mediana) de la conducta de acoso

Ítem	ID	% de personas en la muestra		
		Total[a]	Sin acoso[b]	Acosados[c]
1. Restricción comunicación	.63	24.5 (2)	17.4 (1)	61.8 (3)
2. Ignorar	.63	20.9 (3)	13.7 (2)	58.9 (3)
3. Interrupciones continuas	.66	23.1 (2)	15.3 (2)	64.3 (3)
4. Trabajos contra los principios	.51	16.0 (2)	10.7 (1)	44.0 (3)
5. Evaluación sesgada	.73	28.9 (2)	21.0 (1.5)	71.0 (3)
6. Inactividad forzada	.46	12.7 (2)	8.8 (2)	33.3 (3)
7. Trabajos absurdos	.61	24.3 (2)	19.0 (2)	52.7 (3)
8. Tareas por debajo de competencia	.69	32.0 (2)	25.7 (2)	65.2 (3)
9. Tareas rutinarias	.68	29.2 (2)	22.8 (2)	63.3 (4)
10. Sobrecarga	.60	18.3 (2)	11.3 (1)	55.6 (3)

11. Tareas peligrosas	.44	10.0 (2)	5.7 (2)	32.4 (2)
12. Impedir seguridad	.44	10.6 (2)	6.6 (2)	31.9 (2)
13. Perjuicios económicos	.36	6.8 (1)	4.1 (1)	20.8 (2)
14. Prohibición comunicación compañeros	.46	11.4 (2)	7.2 (2)	33.3 (3)
15. Minusvaloración desempeño	.68	22.3 (2)	14.3 (1)	64.3 (3)
16. Acusaciones difusas	.71	27.2 (2)	18.6 (1)	72.5 (3)
17. Acusación sistemática	.70	24.0 (2)	15.1 (1)	71.0 (3)
18. Amplificación errores	.73	27.7 (2)	19.4 (1)	71.5 (3)
19. Humillaciones	.61	17.4 (2)	9.4 (1)	59.9 (3)
20. Amenaza uso disciplinario	.54	15.6 (2)	9.6 (1)	47.3 (2)
21. Medidas de aislamiento	.47	9.6 (2)	5.2 (2)	32.9 (2)
22. Distorsión comunicación	.62	17.7 (2)	9.5 (1)	60.9 (3)
23. Desestabilización	.65	20.8 (2)	12.7 (1)	63.8 (3)
24. Menosprecio	.67	21.7 (2)	3.0 (1)	67.6 (3)
25. Burlas	.44	11.0 (2)	7.0 (1)	31.9 (3)
26. Críticas vida personal	.42	8.4 (1)	4.7 (1)	28.5 (2)
27. Amenazas en persona	.45	9.0 (2)	3.7 (1)	36.7 (3)
28. Amenazas por escrito	.16	1.5 (1)	1.0 (1)	3.9 (1.5)
29. Gritos	.63	21.8 (2)	14.6 (1)	59.9 (3)
30. Avasallamiento físico	.22	2.1 (1.5)	1.2 (2)	7.2 (1)
31. Ridiculización	.40	8.7 (2)	4.7 (1)	30.0 (3)
32. Rumores y calumnias	.46	10.6 (2)	5.6 (2)	37.2 (2)
33. Privar de información	.58	17.0 (2)	10.3 (1)	52.2 (3)
34. Limitación de carrera profesional	.53	14.3 (2)	8.4 (1)	45.4 (2)
35. Perjuicio imagen	.48	10.3 (2)	5.3 (1)	36.7 (2)
36. Presión indebida	.69	26.1 (2)	18.0 (1)	69.1 (3)
37. Reducción de plazos	.65	25.2 (2)	17.7 (1)	64.7 (3)
38. Modificación de responsabilidad	.61	20.1 (2)	12.8 (2)	58.9 (3)
39. Desvaloración esfuerzo	.68	21.9 (2)	13.1 (1)	68.1 (3)
40. Desmoralizar	.60	15.0 (3)	7.7 (2)	54.1 (3)
41. Inducir a errores	.49	10.2 (2)	4.8 (2)	38.6 (2)
42. Monitorización perversa	.60	16.8 (2)	10.3 (2)	51.2 (3)
43. Insinuación sexual	.28	5.6 (2)	3.3 (2)	17.9 (3)

^{a}N= 1.303; ^{b}n= 1.096; ^{c}n= 207

Us back to Table 2 it can be seen in the discrimination indices of the items (Spearman correlation) calculated in the total sample. In general, have a high discriminatory power: 53% of the items have correlations higher than .60. The items with lower values (items 28, 30 and 43), these results are confirmed in further analysis, the questionnaire should be eliminated.

Validity

Table 4 shows the Spearman correlation between scale scores on scales assessing Cisneros possible consequences of mobbing. The sign of the correlations is theoretically expected, highlighting the correlation between positive and stress mobbing posttraumatic (NEAP: 44; IGAP: 45; IMAP:

165

40). We also investigated whether there were differences between harassed and harassed in the scores on these scales. Table 5 may be descriptive statistics, together with the results of the Mann-Whitney (two-tailed), showing statistically significant differences in all scales, except Burnout scale, which measures Realization personnel ($Z = -1.75$, $p = .08$).

Dimensionality

Stress values for solutions of dimensions 1, 2, 3 and 4 were, respectively, 0,049, 0,023, 0,014 and 0,011. Following the criterion to find a solution ade quately representing the relation between the items (good fit) so parsimonious (Low-dimensional), and theoretical consistency criteria, has chosen the two-dimensional solution. Exploration graph waste dimensional model made visible the fit of the model to the data. The coordinates of each of the items in the two-dimensional space obtained are shown in Table 6. An examination of the content of the items and the value of its coordinates to conclude that the first dimension corresponds with the area to which they relate bullying behaviors. Can conceivable that dimension as a continuum with the pole labor at one end and the staff at the pole opposite end. Thus, items with positive values in the dimension Scope harassment often describe abusive behavior relating to the sphere personal threats writing or by telephone at home, enslavement physical, sexual innuendo, etc.. Items with values negative behaviors often involve abuse directly linked to work: assignments below competition, enforced idleness, routine tasks, jobs absurd etc. The second dimension relates to the nature of the behavior of abuse, the poles can be defined social blockade / humiliation and coercive / punitive. Items with positive values are social blockade consist behaviors and / or humiliation, sometimes quite subtle in nature, as the hints (item 43).

Although, mostly, social blockade strategies and humiliation usually quite explicit: enforced idleness, ignore, teasing, ridicule, among others. Items with negative values are they concerned punitive or coercive in nature, nothing subtle: being threatened in writing or by telephone, prevent take the necessary security measures, overload, and tasks dangerous prohibition of communication, among others.

Discussion and Conclusions.

It is known that the procedures for assessing the reliability and validity of a questionnaire are determined, in part, the nature and objectives of the questionnaire, and another, metrics and statistical properties of the data on which these procedures apply. Not all types of validity are equally relevant, not all data permits apply any statistical procedure (Muñiz, Fidalgo, and Moreno García-Cueto, forthcoming).

The distributional characteristics of most of the variables analyzed have required the use of nonparametric tests but more appropriate than the classical mean and standard deviation. The results offered by such statistical tools show Cisneros scale as a reliable and valid instrument for assessing the mobbing. More specifically, the reliability of the estimated scale so purist, both by the method of the two halves (96), as

by Cronbach's alpha (97), provide values indicate high internal consistency among the items in the questionnaire, along with a good discriminatory (See discrimination indices in Table 2).

The validity of the scale, or better, according to the latest standards of the American Psychological Association, the validity of inferences made from their scores (Elosúa, 2003; Muñiz, 2004), is based on different evidence. The first, an adequate representation of the behavior of interesting. As the main aim of determining the frequency scale Cisneros behaviors occurring with humiliation, harassment and violence

in the workplace, must have a high validity content, defined as the degree to which the conduct that reflects

the questionnaire constitute a representative sample of the mobbing behavior. The remarkable agreement between the behaviors Cisneros sampled by the questionnaire and those reflected questionnaires in other countries have been prepared therefor (Hoel and Cooper, 2000; Leymann 1992; Quinc, 1999, Zapf, Knorz and Kulla, 1996) is a first indicator its content validity. Data derived from the exploration dimensionality of the test provide empirical evidence additional.

Figure 1 shows the distribution of the items on the space defined by the dimensions: Scope of bullying behavior, represented horizontally, and type of bullying behavior, represented vertically. Thus we see graphically how most of the items (42%) are in the left upper quadrant (workplace / type social blockade-humiliation). Moreover, empirically checked how frequent bullying behaviors (see Table 2) correspond

to items located in the quadrant above. This supports the validity of the content of Cisneros scale because, as

stated in the literature, the most common forms of abuse are less coarse and are mostly referred to the field labor (Gonzalez de Rivera and Rodriguez-Abuín, 2002; Leymann, 1996a; Piñuel, 2001, Piñuel and Oñate, 2002).

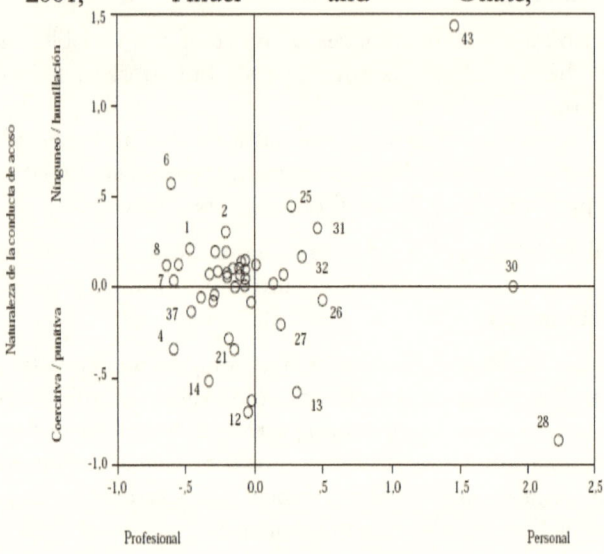

Figura 1. Posición de los ítems en el espacio bidimensional. Por razones de escala sólo se ha podido señalar el número correspondiente a parte de los ítems

Content validity is a necessary but not sufficient to ensure practical, diagnostic scale Cisneros. To this end, should be able to discriminate sufficiently mobbing situations other than those in which interpersonal conflicts occur punctual type. By conventional approach has been adopted, widely accepted by European specialists, with a minimum duration of bullying strategies than six months to be considered that we in a situation of bullying (Einarsen and Skogstad, 1996; Leymann 1996b). The barometer CISNEROS ® includes a question Cisneros scale after this especially, whether the respondent situation has undergone at least for a period six months. Adopting a positive response as a criterion to divide the sample between mobbing sufferers (n =207) and those who do not suffer (n = 1,096), there was a notable difference on the various indices of harassment between the two subsamples (see Table 3) and symptomatology derivative (see Table 5). These results corroborate the validity of both persistence criterion six months, depending on which are set groups, as the sensitivity of the scale to assess Cisneros mobbing. It is symptomatic that Gonzalez de Rivera and Rodriguez-Abuín (2003), applying the LIPT-60 to a sample of 79 people affected by experiences of harassment, indicate an average of behaviors harassment (29 of 60, 48.3%) similar to that found by us in subjects classified as harassment (20 behaviors of 43, 46.5%). However, it would be the study of clinical samples, namely, persons with a clinical diagnosis of mobbing, to characterize precisely to this group.

Ítem	Ámbito de la conducta de acoso	Tipo de conducta de acoso
Tabla 6 Coordenadas de los ítems en el espacio bidimensional		
Dimensión		
1. Restricción comunicación	-0.469	0.210
2. Ignorar	-0.210	0.301
3. Interrupciones continuas	-0.288	0.194
4. Trabajos contra los principios	-0.595	-0.344
5. Evaluación sesgada	-0.322	0.073
6. Inactividad forzada	-0.615	0.571
7. Trabajos absurdos	-0.596	0.027
8. Tareas por debajo de competencia	-0.647	0.113
9. Tareas rutinarias	-0.551	0.114
10. Sobrecarga	-0.289	0.048
11. Tareas peligrosas	-0.019	-0.636
12. Impedir seguridad	-0.051	-0.704
13. Perjuicios económicos	0.316	-0.592
14. Prohibición comunicación compañeros	-0.330	-0.534
15. Minusvaloración desempeño	-0.199	0.069
16. Acusaciones difusas	-0.262	0.074
17. Acusación sistemática	-0.158	0.092

18. Amplificación errores	-0.195	0.056
19. Humillaciones	-0.063	0.141
20. Amenaza uso disciplinario	-0.175	-0.293
21. Medidas de aislamiento	-0.147	-0.345
22. Distorsión comunicación	-0.063	0.003
23. Desestabilización	-0.061	0.086
24. Menosprecio	-0.092	0.120
25. Burlas	0.274	0.439
26. Críticas vida personal	0.501	-0.085
27. Amenazas en persona	0.202	-0.218
28. Amenazas por escrito	2.229	-0.864
29. Gritos	-0.199	0.185
30. Avasallamiento físico	1.889	-0.015
31. Ridiculización	0.469	0.316
32. Rumores y calumnias	0.353	0.162
33. Privar de información	-0.062	0.048
34. Limitación de carrera profesional	-0.023	-0.091
35. Perjuicio imagen	0.217	0.051
36. Presión indebida	-0.297	-0.083
37. Reducción de plazos	-0.465	-0.144
38. Modificación de responsabilidad	-0.391	-0.065
39. Desvaloración esfuerzo	-0.141	-0.005
40. Desmoralizar	0.017	0.113
41. Inducir a errores	0.145	0.015
42. Monitorización perversa	-0.100	0.056
43. Insinuación sexual	1.461	1.436

According to the literature (Baron, Munduate and White, 2003; Hoel and Cooper, 2000; Leymann and Gustaffson, 1996; Niedl, 1996; Piñuel, 2001, 2003; Piñuel and Oñate, 2002), and psychological health physical persons suffering mobbing often suffer a great deterioration. Therefore, it is expected that the scale scores Cisneros, as a good indicator of psychological harassment, pertaining to the measuring instruments of this series alterations. The results fit the expectations theoretical, supporting the construct validity of the scale

(See Tables 4 and 5). Highlighting the high correlation that Cisneros has with Posttraumatic Stress Scale (48),

confirming the results of investigations of Leymann and Gustafsson (1996), indicating that the more psychological diagnosis frequent among those workers victimized by mobbing is to PTSD. It is also confirmed depression, according to other researchers in the field (Zapf et al., 1996) as a sequela specific victims mobbing, with a correlation between the scale and Cisneros Beck scale of 29 and a difference of 5 points on the scale

Beck among bullied and harassed. The intention to leave the Company is characteristic of a problem like that mobbing settles in over 80% of cases, with the output of the worker organization (Niedl, 1996; Piñuel, 2001). In this regard deserves is worth noting the correlation between scores Cisneros on the scale and the scale of Abandonment (39), and high incidence rate of intent to leave the profession mobbing victims have a difference of 4 points on the average score of non harassed. Similarly, two scales Burnout (emotional exhaustion and depersonalization) correlations exhibit substantial and statistically significant differences between samples bullied and harassed. All these data emphasize the need to differentiate cases of psychological violence from those in that this does not occur, although there similar damage, especially the need for differential diagnosis between mobbing and burnout (I Boada Grau, Agulló Diego Vallejo and Thomas, 2004; Piñuel, 2003).

Finally, we offer guidelines that help interpret scores on the scale Cisneros. Correction scale allows for three indices that report: first, the number of strategies suffered harassment (NEAP), with a range 0-43 theorist, and second, the overall magnitude of harassment (IGAP), with a theoretical range of 0 (zero) to 6 (maximum), and third, average intensity behaviors suffered harassment (IMAP), with a theoretical range from 0 (zero) and 6 (maximum). Also the theoretical range of the indices in order to have a point to reference to interpret the scores in Table 7. The values corresponding to the quartiles of the distribution of the three indices calculated on bullying total sample of people without bullying and harassment. We can conclude, on the basis of the results, that Cisneros is a reliable assessment tool, with proper sampling of the contents, and high internal consistency shows relationships with variables of interest that fit the pattern theoretical expected. On the other hand, the breadth and diversity of the shows, unusual in Spanish research on mobbing, and much information that further use of this instrument has provided (Piñuel, 2001b; Piñuel and Oñate, 2002), make it a reference scale for assessing the mobbing. It is especially relevant shows the high correlation in relation the most common manifestations of damage in the mobbing, especially in regard to the pictures of PTSD and depression. This can be especially useful for damage assessment presented those victims in treatment psychotherapy and also as a diagnostic tool expert opinion in court cases increasingly frequent in of mobbing. Finally, we point out possible research that should be undertaken, some already underway. First, determine the factor structure of the scale and comparative analysis Similar assessment tools (Piñuel et al. 2004a). Second, to complement the results presented those obtained in samples from individuals with a clinical diagnosis of mobbing. Third, systematically investigate the meaning and predictive utility of the various indices of harassment. And fourth, replicate the study and the questionnaire applied to new populations, allowing for a more accurate baremación (Piñuel et al. 2004b).

Tabla 7

Puntuaciones en los índices de acoso correspondientes a los cuartiles de distribución

	Muestra								
	Total[a]			Sin acoso[b]			Acosados[c]		
Percentil	NEAP	IGAP	IMAP	NEAP	IGAP	IMAP	NEAP	IGAP	IMAP
25	0	0.00	0.00	0	0.00	0.00	13	0.67	2.14
50	2	0.07	1.00	0	0.00	0.00	20	1.23	2.80
75	12	0.51	2.00	7	0.26	1.40	29	2.19	3.79

NEAP: Número de estrategias de acoso. IGAP: Índice global del acoso. IMAP: Intensidad promedio del acoso.
[a]$N= 1.303$. [b]$n= 1.096$. [c]$n= 207$.

¿Cuáles de las siguientes formas de maltrato psicológico (ver lista de preguntas 1 a 43) se han ejercido contra Ud.?

Señale, en su caso, quién/es son el/los autor/es de los hostigamientos recibidos

1   Jefes o supervisores
2   Compañeros de trabajo
3   Subordinados

Señale, en su caso, el grado de frecuencia con que se producen esos hostigamientos

0   Nunca
1   Pocas veces al año o menos
2   Una vez al mes o menos
3   Algunas veces al mes
4   Una vez a la semana
5   Varias veces a la semana
6   Todos los días

Comportamientos	Autor/es	Frecuencia del comportamiento						
1. Mi superior restringe mis posibilidades de comunicarme, hablar o reunirme con él	| |	0	1	2	3	4	5	6
2. Me ignoran, me excluyen o me hacen el vacío, fingen no verme o me hacen «invisible»	| |	0	1	2	3	4	5	6
3. Me interrumpen continuamente impidiendo expresarme	| |	0	1	2	3	4	5	6
4. Me fuerzan a realizar trabajos que van contra mis principios o mi ética	| |	0	1	2	3	4	5	6
5. Evalúan mi trabajo de manera inequitativa o de forma sesgada	| |	0	1	2	3	4	5	6
6. Me dejan sin ningún trabajo que hacer, ni siquiera a iniciativa propia	| |	0	1	2	3	4	5	6
7. Me asignan tareas o trabajos absurdos o sin sentido	| |	0	1	2	3	4	5	6
8. Me asignan tareas o trabajos por debajo de mi capacidad profesional o mis competencias	| |	0	1	2	3	4	5	6
9. Me asignan tareas rutinarias o sin valor o interés alguno	| |	0	1	2	3	4	5	6
10. Me abruman con una carga de trabajo insoportable de manera malintencionada	| |	0	1	2	3	4	5	6
11. Me asignan tareas que ponen en peligro mi integridad física o mi salud a propósito	| |	0	1	2	3	4	5	6
12. Me impiden que adopte las medidas de seguridad necesarias para realizar mi trabajo con la debida seguridad	| |	0	1	2	3	4	5	6
13. Se me ocasionan gastos con intención de perjudicarme económicamente	| |	0	1	2	3	4	5	6
14. Prohíben a mis compañeros o colegas hablar conmigo	| |	0	1	2	3	4	5	6
15. Minusvaloran y echan por tierra mi trabajo, no importa lo que haga	| |	0	1	2	3	4	5	6
16. Me acusan injustificadamente de incumplimientos, errores, fallos, inconcretos y difusos	| |	0	1	2	3	4	5	6
17. Recibo críticas y reproches por cualquier cosa que haga o decisión que tome en mi trabajo	| |	0	1	2	3	4	5	6
18. Se amplifican y dramatizan de manera injustificada errores pequeños o intrascendentes	| |	0	1	2	3	4	5	6
19. Me humillan, desprecian o minusvaloran en público ante otros colegas o ante terceros	| |	0	1	2	3	4	5	6
20. Me amenazan con usar instrumentos disciplinarios (rescisión de contrato, expedientes, despido, traslados, etc.)	| |	0	1	2	3	4	5	6
21. Intentan aislarme de mis compañeros dándome trabajos o tareas que me alejan físicamente de ellos	| |	0	1	2	3	4	5	6
22. Distorsionan malintencionadamente lo que digo o hago en mi trabajo	| |	0	1	2	3	4	5	6
23. Se intenta buscarme las cosquillas para «hacerme explotar»	| |	0	1	2	3	4	5	6
24. Me menosprecian personal o profesionalmente	| |	0	1	2	3	4	5	6
25. Hacen burla de mí o bromas intentando ridiculizar mi forma de hablar, de andar, etc.	| |	0	1	2	3	4	5	6
26. Recibo feroces e injustas críticas acerca de aspectos de mi vida personal	| |	0	1	2	3	4	5	6
27. Recibo amenazas verbales o mediante gestos intimidatorios	| |	0	1	2	3	4	5	6
28. Recibo amenazas por escrito o por teléfono en mi domicilio	| |	0	1	2	3	4	5	6
29. Me chillan o gritan, o elevan la voz de manera a intimidarme	| |	0	1	2	3	4	5	6
30. Me zarandean, empujan o avasallan físicamente para intimidarme	| |	0	1	2	3	4	5	6
31. Se hacen bromas inapropiadas y crueles acerca de mí	| |	0	1	2	3	4	5	6
32. Inventan y difunden rumores y calumnias acerca de mí de manera malintencionada	| |	0	1	2	3	4	5	6
33. Me privan de información imprescindible y necesaria para hacer mi trabajo	| |	0	1	2	3	4	5	6
34. Limitan malintencionadamente mi acceso a cursos, promociones, ascensos, etc.	| |	0	1	2	3	4	5	6
35. Me atribuyen malintencionadamente conductas ilícitas o antiéticas para perjudicar mi imagen y reputación	| |	0	1	2	3	4	5	6
36. Recibo una presión indebida para sacar adelante el trabajo	| |	0	1	2	3	4	5	6
37. Me asignan plazos de ejecución o cargas de trabajo irrazonables	| |	0	1	2	3	4	5	6

			0	1	2	3	4	5	6
38. Modifican mis responsabilidades o las tareas a ejecutar sin decirme nada	[ ]		0	1	2	3	4	5	6
39. Desvaloran continuamente mi esfuerzo profesional	[ ]		0	1	2	3	4	5	6
40. Intentan persistentemente desmoralizarme	[ ]		0	1	2	3	4	5	6
41. Utilizan varias formas de hacerme incurrir en errores profesionales de manera malintencionada	[ ]		0	1	2	3	4	5	6
42. Controlan aspectos de mi trabajo de forma malintencionada para intentar «pillarme en algún renuncio»	[ ]		0	1	2	3	4	5	6
43. Me lanzan insinuaciones o proposiciones sexuales directas o indirectas	[ ]		0	1	2	3	4	5	6
44. En el transcurso de los últimos 6 meses, ¿ha sido Ud víctima de por lo menos alguna de las anteriores formas de maltrato psicológico de manera continuada (con una frecuencia de más de 1 vez por semana)? (ver lista de preguntas la 43)	☐ si ☐ no								

References

Barón, M., Munduate, L. y Blanco, M.J. (2003). La espiral del mobbing.*Papeles del Psicólogo, 84,* 71-82.

Beck, A.T., Rush, A.J., Shaw, B.F. y Emery, G. (1983). *Terapia Cognitiva de la Depresión.* Bilbao: Desclée de Brower.

Boada i Grau, J., De Diego Vallejo, R. y Agulló Tomás, E. (2004). El burnout y las manifestaciones psicosomáticas como consecuentes del clima organizacional y de la motivación laboral. *Psicothema, 16,* 125-131.

Coxon, A.P.M. (1982). *The User's Guide to multidimensional scaling.* London: Heinemann Educational Books.

Curbow, B. y Somerfield, M. (1991). Use of Rosenberg Self-Esteem Scale with Adult Cancer Patients. *Journal of Psychosocial Oncology, 9* (2), 113-131.

Davenport, N., Schwartz, R.D. y Elliot, G.P. (1999). *Mobbing. Emocional abuse in the american workplace.* Ames, Iowa: Civil Society Publishing.

Einarsen, S. y Skogstad, A. (1996). Bullying at work. Epidemiological findings in public and private organizations. *European Journal of Work and Organizational Pychology, 5,* 185-203.

Elosúa, P. (2003). Sobre la validez de los test. *Psicothema, 15,* 315-321.

González de Rivera, J.L. y Rodríguez-Abuín, M. (2003). Cuestionario de estrategias de acoso psicológico: el LIPT-60 (Leymann Inventory of Psychological Terrorization) en versión española. *Psiquis, 24* (2), 59-66.

Heinemann, P. (1972). *Mobbing-Group violence by children and adults.*Stockholm: Natur och Kultur.

Hirigoyen, M.F. (1994). *Le harcèlement morale.* París: Syros.

Hirigoyen, M.F. (2001). *El acoso moral.* Barcelona: Paidós.

Hoel, H. y Cooper, G. (2000). *Destructive conflict and bullying at work.*Unpublished report. University of Manchester.

Leymann, H. (1986). *Mobbing-Psychological violence at work places.* Lund: Studentlitteratur.

Leymann, H. (1990). Mobbing and psychological terror at wokplaces. *Violence and victims, 5,* 119-126.

Leymann, H. (1992). *Leymann inventory of psychological terror.* Violen: Karlskrona.

Leymann, H. (1993). *Mobbing. Psychoterror am Arbeitsplatz und wie man sich dagegen wehrenkann.* Reinbek: Rowohlt.

Leymann, H. (1996a). *Mobbing. La persécution au travail.* Paris: Seuil.

Leymann, H. (1996b). The content and development of mobbing at work *The European Journal of Work and Organizational Pychology, 5,* 165-184.

Leymann, H. y Gustafsson, A. (1996). Mobbing at work and the development of Post-traumatic Stress Disorders. *The European Journal of Work and Organizational Pychology, 5,* 251-277.

Lorenz, K. (1991). *Hier bin ich-wo bist Du? Ethologie der grauganz.* München: Piper.

Maslach, C. y Jackson, S.E. (1997). *MBI inventario «Burnout» de Maslach.* Madrid: TEA.

Muñiz, J. (2004). La validación de los tests. *Metodología de las Ciencias del Comportamiento, 5,* 121-141.

Muñiz, J., Fidalgo, A.M., García-Cueto, E. y Moreno, R. (en prensa). *Análisis de los ítems de cuestionarios y tests.* Madrid: La Muralla/Hespérides.

Niedl, K. (1996). Mobbing and well being: economic and personnel development implications. *European Journal of Work and Organizational Pychology, 5,* 239-249.

Piñuel, I. (2001a). *Mobbing. Cómo sobrevivir al acoso psicológico en el trabajo.* Santander: Sal terrae.

Piñuel, I. (2001b). Mobbing. La lenta y silenciosa alternativa al despido. *AEDIPE, 17,* 19-55.

Piñuel, I. (2003). *Mobbing. Manual de autoayuda.* Madrid: Aguilar.

Piñuel, I., Fidalgo, A.M., Oñate, A. y Ferreres, D. (2004a, abril). *Dimensiones y factores del mobbing o acoso psicológico en el trabajo en España.Análisis comparativo con otros estudios europeos.* Comunicación presentada a la VII European Conference on Psychological

Assessment, Málaga.

Piñuel, I., Fidalgo, A.M., Oñate, A. y Ferreres, D. (2004b, abril). *Resultados epidemiológicos del barómetro Cisneros III sobre acoso psicológico en el trabajo o Moobing.* Comunicación presentada a la VII European Conference on Psychological Assessment, Málaga.

Piñuel, I. y Oñate, A. (2002). La incidencia del mobbing o acoso psicológico en el trabajo en España. *Lan Harremanak, 7*(II), 35-62.

Piñuel, I. y Oñate, A. (2003). El mobbing o acoso psicológico en el trabajo en España. *Congreso Internacional Virtual: intangibles e interdisciplinariedad,*409-426.

Quinc, L. (1999). Workplace bulling in NHS community trust: staff questionnaire survey. *British Medical Journal, 318,* 228-232.

Real Deus, J.E. (2001). Escalamiento multidimensional. Madrid: La Muralla/Hespérides.

Rosenberg, M. (1965). *Society and the Adolescent Self Image.* Princenton, N.J.: Princenton University Press.

Zapf, D., Knorz, C. y Kulla, M. (1996). On the relationship between Mobbing factors and Job Content. *European Journal of Work and Organizational Pychology, 5,* 215-239.

Zapf, D. y Einersen, S. (Eds.) (2001). Bulling in the work-place: Recent trends in research and practice (special issue). *European Journal of Work and Organizational Pychology, 10*(4).

www.ingramcontent.com/pod-product-compliance
Lightning Source LLC
Chambersburg PA
CBHW032015170526
45157CB00002B/703